The IMA Volumes in Mathematics and its Applications

Volume 160

Series editor

Fadil Santosa, *University of Minnesota, MN, USA*

Institute for Mathematics and its Applications (IMA)

The Institute for Mathematics and its Applications (IMA) was established in 1982 as a result of a National Science Foundation competition. The mission of the IMA is to connect scientists, engineers, and mathematicians in order to address scientific and technological challenges in a collaborative, engaging environment, developing transformative, new mathematics and exploring its applications, while training the next generation of researchers and educators. To this end the IMA organizes a wide variety of programs, ranging from short intense workshops in areas of exceptional interest and opportunity to extensive thematic programs lasting nine months. The IMA Volumes are used to disseminate results of these programs to the broader scientific community.

The full list of IMA books can be found at the Web site of the Institute for Mathematics and its Applications:

> http://www.ima.umn.edu/springer/volumes.html.

Presentation materials from the IMA talks are available at

> http://www.ima.umn.edu/talks/.

Video library is at

> http://www.ima.umn.edu/videos/.

Fadil Santosa, Director of the IMA

More information about this series at http://www.springer.com/series/811

Susanne C. Brenner
Editor

Topics in Numerical Partial Differential Equations and Scientific Computing

ASSOCIATION FOR
WOMEN IN MATHEMATICS

Editor
Susanne C. Brenner
Department of Mathematics
  and Center for Computation and Technology
Louisana State University
Baton Rouge, LA
USA

ISSN 0940-6573   ISSN 2198-3224 (electronic)
The IMA Volumes in Mathematics and its Applications
ISBN 978-1-4939-8187-8   ISBN 978-1-4939-6399-7   (eBook)
DOI 10.1007/978-1-4939-6399-7

Mathematics Subject Classification (2010): 65

Foreword

This volume is based on the research initiated at the IMA's WhAM! A Research Collaboration Workshop for Women in Applied Mathematics that took place in August 12–15, 2014. This was the second of such workshops and focused on numerical partial differential equations and scientific computing. The goals of this workshop series are to foster the formation of new research groups and to create a supportive network of women researchers working in the area of numerical analysis and scientific computing.

The IMA owes a lot of gratitude to Susanne C. Brenner for taking a leading role in organizing this particular workshop. She not only recruited excellent senior researchers who brought projects to work on, but she was also active in recruiting graduate students and postdocs to participate. She should be commended for the mentoring she provided to the younger participants during the entire week. Without her hard work during and after the workshop, this volume would not have been possible.

By any measure, the workshop was successful in accomplishing its goals, but the most tangible result from the workshop is this book. Each article is based on original research and reports on the findings of the teams.

The IMA thanks Sigal Gottlieb, Chiu-Yen Kao, Hyesuk Lee, Fengyan Li, and Carol Woodward for coorganizing the workshop and for bringing such interesting projects. Finally, we thank the National Science Foundation for its support of the IMA.

Minneapolis, MN, USA   Fadil Santosa

Contents

Introduction

The second WhAM! research collaboration workshop for women in applied mathematics took place at the Institute for Mathematics and its Applications (IMA) in Minneapolis during August 12–15, 2014. The focus of this workshop was on numerical partial differential equations and scientific computing. There were thirty-three participants composed of graduate students, postdocs, and mid-career and senior researchers. They worked in six teams on projects in numerical algorithms and their applications. The collaborations initiated in this workshop continued throughout the following year, during which each team held a follow-up meeting to strengthen and consolidate the research efforts. The final results are presented in the six peer-reviewed chapters of this volume.

The chapter "A C^0 Interior Penalty Method for Elliptic Distributed Optimal Control Problems in Three Dimensions with Pointwise State Constraints" is the work of the team led by Susanne C. Brenner and Natasha S. Sharma, with team members Minah Oh, Sara Pollock, Kamana Porwal, and Mira Schedensack.

The chapter "The Effect of the Sensitivity Parameter in Weighted Essentially Non-oscillatory Methods" is the work of the team led by Sigal Gottlieb and Bo Dong, with team members Yulia Hristova, Yan Jiang, and Haijin Wang.

The chapter "Study of a Mixed Dispersal Population Dynamics Model" is the work of the team led by Chiu-Yen Kao and Marina Chugunova, with team members Baasansuren Jadamba, Christine Klymko, Evelyn Thomas, and Bingyu Zhao.

The chapter "Optimization-Based Decoupling Algorithms for a Fluid-Poroelastic System" is the work of the team led by Hyesuk Lee and Annalisa Quaini, with team members Aycil Cesmelioglu, Kening Wang, and Son-Young Yi.

The chapter "Study of Discrete Scattering Operators for Some Linear Kinetic Models" is the work of the team led by Fengyan Li and Yingda Cheng, with team members Yanping Chen, Zheng Chen, and Adrianna Gillman.

The chapter "On Metrics for Computation of Strength of Coupling in Multiphysics Simulations" is the work of the team led by Carol S. Woodward and Yekaterina Epshteyn, with team members Anastasia Wilson, Wei Du, Guanglian Li, and Azam Moosavi.

The opportunities for research, mentoring, and networking provided by this program are greatly appreciated by the participants, who are grateful to the IMA for their hospitality and the National Science Foundation and Microsoft for their financial support.

A C^0 Interior Penalty Method for Elliptic Distributed Optimal Control Problems in Three Dimensions with Pointwise State Constraints

Susanne C. Brenner, Minah Oh, Sara Pollock, Kamana Porwal, Mira Schedensack and Natasha S. Sharma

Abstract We investigate numerically a triquadratic C^0 interior penalty method for elliptic distributed optimal control problems in three dimensions with pointwise state constraints, which is based on the formulation of these problems as fourth order variational inequalities. We obtain numerical results that are similar to the ones reported in [7, 8] for fourth order variational inequalities in two dimensions. The deal.II library [1, 2] is used for the numerical experiments.

S.C. Brenner (✉) · K. Porwal
Department of Mathematics and Center for Computation and Technology,
Louisiana State University, Baton Rouge, LA 70803, USA
e-mail: brenner@math.lsu.edu

K. Porwal
e-mail: kporwal@math.lsu.edu

M. Oh
Department of Mathematics and Statistics, James Madison University,
Harrisonburg, VA 22807, USA
e-mail: ohmx@jmu.edu

S. Pollock
Department of Mathematics, Texas A&M University, College Station, TX 77843, USA
e-mail: snpolloc@math.tamu.edu

M. Schedensack
Institut für Numerische Simulation, Universität Bonn,
Wegelerstraße 6, 53115 Bonn, Germany
e-mail: schedensack@ins.uni-bonn.de

N.S. Sharma
Department of Mathematical Sciences,
University of Texas of El Paso, El Paso, TX 79968, USA
e-mail: nssharma@utep.edu

© Springer Science+Business Media New York 2016
S.C. Brenner (ed.), *Topics in Numerical Partial Differential Equations and Scientific Computing*, The IMA Volumes in Mathematics and its Applications 160, DOI 10.1007/978-1-4939-6399-7_1

1

1 Introduction

We consider a class of optimal control problems with pointwise state constraints over a bounded convex polyhedral domain $\Omega \subset \mathbb{R}^3$. We first recall the standard notation and introduce the functional setting for both the optimal control problems and their characterizations as fourth order variational inequalities. The space $L^2(\Omega)$ denotes the space of square integrable functions on Ω, and $L_0^2(\Omega)$ is the space of functions in $L^2(\Omega)$ with zero mean. We use $H^s(\Omega)$ to denote the set of all $L^2(\Omega)$ functions whose distributional derivatives up to order s are in $L^2(\Omega)$, and $H_0^s(\Omega)$ to denote the set of functions in $H^s(\Omega)$ whose traces vanish up to order $s-1$ on $\partial\Omega$. The corresponding inner product and norm defined on these Hilbert spaces will be denoted by $(\cdot,\cdot)$ and $\|\cdot\|$, respectively, with the function space as a subscript, i.e., $(\cdot,\cdot)_{H^1}$ and $\|\cdot\|_{H^1}$, etc., and similarly for the seminorm $|\cdot|$. We will sometimes omit the subscript in the case of the inner product and norm of $L^2(\Omega)$.

For $\psi \in C^2(\bar\Omega)$, $y_d \in L^2(\Omega)$, and γ a positive constant, we define the sets $\mathcal{K}_D \subset H_0^1(\Omega) \times L^2(\Omega)$ and $\mathcal{K}_N \subset H^1(\Omega) \times L_0^2(\Omega)$ by

$$
\mathcal{K}_D = \{(y,u) \in H_0^1(\Omega) \times L^2(\Omega) :
$$
$$
\int_\Omega \nabla y \cdot \nabla z\, dx = \int_\Omega uz\, dx \text{ for all } z \in H_0^1(\Omega) \text{ and } y \le \psi \text{ a.e. in } \Omega\},
$$
$$
\mathcal{K}_N = \{(y,u) \in H^1(\Omega) \times L_0^2(\Omega) :
$$
$$
\int_\Omega \nabla y \cdot \nabla z\, dx = \int_\Omega uz\, dx \text{ for all } z \in H^1(\Omega) \text{ and } y \le \psi \text{ a.e. in } \Omega\}.
$$

We will consider the following elliptic distributed optimal control problem:

$$
\text{Find } (\bar y, \bar u) = \underset{(y,u)\in\mathcal{K}}{\operatorname{argmin}} \left[\frac{1}{2}\|y - y_d\|^2 + \frac{\gamma}{2}\|u\|^2\right], \tag{1}
$$

where $\mathcal{K} = \mathcal{K}_D$ (Dirichlet problem) or $\mathcal{K} = \mathcal{K}_N$ (Neumann problem).

Let the spaces V_D and V_N be defined by

$$
V_D = H^2(\Omega) \cap H_0^1(\Omega) \quad \text{and} \quad V_N = \left\{v \in H^2(\Omega) : \frac{\partial v}{\partial n} = 0 \text{ on } \partial\Omega\right\}.
$$

Since $(y,u) \in \mathcal{K}$ implies $y \in H^2(\Omega)$ and $u = -\Delta y$ by elliptic regularity [15], the optimal control problem (1) is equivalent to the following problem:

$$
\text{Find } \bar y = \underset{y\in K}{\operatorname{argmin}} \left[\frac{1}{2}\|y - y_d\|^2 + \frac{\gamma}{2}\|\Delta y\|^2\right] \tag{2}
$$
$$
= \underset{y\in K}{\operatorname{argmin}} \left[\frac{\gamma}{2}\|\Delta y\|^2 + \frac{1}{2}\|y\|^2 - (y_d, y)\right],
$$

where

$$K = K_D = \{y \in V_D : y \leq \psi \text{ in } \Omega\} \qquad \text{(Dirichlet problem)},$$

or

$$K = K_N = \{y \in V_N : y \leq \psi \text{ in } \Omega\} \qquad \text{(Neumann problem)}.$$

Let $D^2 y : D^2 z$ denote the (Frobenius) inner product of the Hessian matrices of y and z:

$$D^2 y : D^2 z = \sum_{1 \leq i,j \leq 3} \left(\frac{\partial^2 y}{\partial x_i \partial x_j} \right) \left(\frac{\partial^2 z}{\partial x_i \partial x_j} \right).$$

It follows from integration by parts that (2) can be rewritten as

$$\text{Find } \bar{y} = \underset{y \in K}{\operatorname{argmin}} \left[\frac{1}{2} \mathcal{A}(y, y) - (y_d, y) \right], \tag{3}$$

where

$$\mathcal{A}(y, z) = \gamma \int_{\Omega} D^2 y : D^2 z \, dx + \int_{\Omega} yz \, dx. \tag{4}$$

Note that in the Neumann case the closed convex subset K_N of V_N is always non-empty since it contains all constant functions that are bounded above by $\min_{x \in \Omega} \psi(x)$. On the other hand, in the Dirichlet case we assume that $\psi > 0$ on $\partial\Omega$ so that K_D is a nonempty subset of V_D and the contact set where $y = \psi$ is disjoint from $\partial\Omega$. Since $\mathcal{A}(\cdot, \cdot)$ is symmetric, bounded, and coercive on $H^2(\Omega)$, the standard theory [13, 22, 23, 26] implies that (3) has a unique solution characterized by the variational inequality

$$\mathcal{A}(\bar{y}, y - \bar{y}) \geq (f, y - \bar{y}) \text{ for all } y \in K,$$

where $K = K_D$ or $K = K_N$.

The goal of this paper is to demonstrate that C^0 interior penalty methods [3–9, 11, 16] are effective for the numerical solution of (3). We note that in the literature, the Dirichlet problem defined by (1) is solved as a fourth order variational inequality by a Morley finite element method in [24], a mixed finite element method in [14], and a quadratic C^0 interior penalty method in [8, 9]. However, the numerical examples in these references only involve two-dimensional domains. To the best of our knowl-edge, this is the first paper that provides numerical results for elliptic distributed optimal control problems in three dimensions formulated as fourth order variational inequalities.

The rest of the paper is organized as follows. In Section 2, we introduce the discrete problems for (3) that are based on the C^0 interior penalty approach. We further discuss three procedures that generate approximations of the optimal control $\bar{u}$ by post-processing the discrete optimal state. In Section 3, we will provide some details concerning the implementation of a primal-dual active set method using the deal.II library. In Section 4, which is the main section, we present numerical results for both the Dirichlet problem and the Neumann problem. Finally, we end with some concluding remarks in Section 5.

2 Discrete Problems

Let $\mathscr{T}_h$ be a uniform triangulation of Ω by cubic elements, $V_h \subset H^1(\Omega)$ be the (continuous) $\mathbb{Q}_2$ finite element space associated with $\mathscr{T}_h$, and let $\mathring{V}_h \subset H_0^1(\Omega)$ be the subspace of V_h whose members vanish on $\partial\Omega$. We will use the following notation throughout the paper:

- h is a mesh parameter proportional to $\max_{T \in \mathscr{T}_h} \operatorname{diam} T$.
- h_F is the diameter of the face F.
- $\mathcal{V}_h$ is the set of vertices of $\mathscr{T}_h$.
- $\mathcal{F}_h$ is the set of faces of $\mathscr{T}_h$.
- $\mathcal{F}_h^i$ is the set of interior faces of Ω.

Let $F \in \mathcal{F}_h^i$ be the common face of $T_\pm \in \mathscr{T}_h$ and n_F be the unit normal of F pointing from T_- to T_+. The jump $[\![\cdot]\!]$ and average $\{\!\!\{\cdot\}\!\!\}$ of the normal derivatives over F for functions in the piecewise Sobolev spaces

$$H^s(\Omega, \mathscr{T}_h) = \left\{ v \in L^2(\Omega) : v_T = v|_T \in H^s(T) \text{ for all } T \in \mathscr{T}_h \right\}$$

are defined as follows:

$$\left\{\!\!\left\{ \frac{\partial^2 v}{\partial n^2} \right\}\!\!\right\} = \frac{1}{2}\left(\frac{\partial^2 v_+}{\partial n_F^2}\Big|_F + \frac{\partial^2 v_-}{\partial n_F^2}\Big|_F \right) \quad \text{for all } v \in H^s(\Omega, \mathscr{T}_h),\ s > \frac{5}{2},$$

$$\left[\!\!\left[\frac{\partial v}{\partial n} \right]\!\!\right] = \frac{\partial v_+}{\partial n_F}\Big|_F - \frac{\partial v_-}{\partial n_F}\Big|_F \quad \text{for all } v \in H^s(\Omega, \mathscr{T}_h),\ s > \frac{3}{2},$$

where $v_\pm = v|_{T_\pm}$.

For $F \in \mathcal{F}_h$ that is a subset of $\partial\Omega$, the jump and average are defined by

$$\left[\!\!\left[\frac{\partial v}{\partial n} \right]\!\!\right] = -\frac{\partial v}{\partial n_F} \quad \text{and} \quad \left\{\!\!\left\{ \frac{\partial^2 v}{\partial n^2} \right\}\!\!\right\} = \frac{\partial^2 v}{\partial n_F^2},$$

where n_F is the unit normal of F pointing towards the outside of Ω.

2.1 Dirichlet Problem

Let the closed convex subset $K_{D,h} \subset \mathring{V}_h$ be defined by

$$K_{D,h} = \{y \in \mathring{V}_h : y(p) \le \psi(p) \text{ for all } p \in V_h\}.$$

The discrete problem for (3) when $K = K_D$ then reads:

$$\text{Find } \bar{y}_h = \underset{y_h \in K_{D,h}}{\text{argmin}} \left[\frac{1}{2}\mathcal{A}_{D,h}(y_h, y_h) - (f, y_h)\right], \tag{5}$$

where

$$\mathcal{A}_{D,h}(v, w) = \gamma a_{D,h}(v, w) + (v, w)$$

and

$$a_{D,h}(v, w) = \sum_{T \in \mathcal{T}_h} \int_T D^2 v : D^2 w \, dx + \sum_{F \in \mathcal{F}_h^i} \int_F \{\!\!\{\partial^2 v/\partial n^2\}\!\!\} [\![\partial w/\partial n]\!] dS$$

$$+ \sum_{F \in \mathcal{F}_h^i} \int_F \{\!\!\{\partial^2 w/\partial n^2\}\!\!\} [\![\partial v/\partial n]\!] dS \tag{6}$$

$$+ \sigma \sum_{F \in \mathcal{F}_h^i} h_F^{-1} \int_F [\![\partial v/\partial n]\!] [\![\partial w/\partial n]\!] dS.$$

Here, $\sigma > 0$ is a penalty parameter chosen large enough (cf. [20]) so that $a_{D,h}(\cdot, \cdot)$ is positive definite on V_h. Note that the sums in (6) involving the jumps and the averages run only over the interior faces.

Remark 1 The finite element space $\mathring{V}_h$ and the bilinear form $a_{D,h}$ appear in C^0 interior penalty methods for the biharmonic equation with the boundary conditions of simply supported plates [3, 5, 11].

2.2 Neumann Problem

Let the closed convex subset $K_{N,h} \subset V_h$ be defined by

$$K_{N,h} = \{y \in V_h : y(p) \le \psi(p) \text{ for all } p \in V_h\}.$$

The discrete problem for (3) when $K = K_N$ is defined as follows.

$$\text{Find } \bar{y}_h = \underset{y_h \in K_{N,h}}{\text{argmin}} \left[\frac{1}{2}\mathcal{A}_{N,h}(y_h, y_h) - (f, y_h)\right], \tag{7}$$

where

$$\mathcal{A}_{N,h}(v, w) = \gamma a_{N,h}(v, w) + (v, w)$$

and

$$
\begin{aligned}
a_{N,h}(v, w) = \sum_{T \in \mathcal{T}_h} \int_T D^2 v : D^2 w \, dx &+ \sum_{F \in \mathcal{F}_h} \int_F \{\!\!\{\partial^2 v / \partial n^2\}\!\!\} \, [\![\partial w / \partial n]\!] \, dS \\
&+ \sum_{F \in \mathcal{F}_h} \int_F \{\!\!\{\partial^2 w / \partial n^2\}\!\!\} \, [\![\partial v / \partial n]\!] \, dS \\
&+ \sigma \sum_{F \in \mathcal{F}_h} h_F^{-1} \int_F [\![\partial v / \partial n]\!][\![\partial w / \partial n]\!] \, dS.
\end{aligned}
\tag{8}
$$

In contrast to (6), the sums in (8) involving the jumps and the averages run over all faces.

Remark 2 The finite element space V_h and the bilinear form $a_{N,h}$ appear in C^0 interior penalty methods for the biharmonic equation with boundary conditions of the Cahn–Hilliard type [3, 4].

2.3   Post-processing

We now describe three post-processing procedures from [9] that generate approximations $\bar{u}_h$ for the optimal control $\bar{u}$ from the discrete optimal state $\bar{y}_h$.

2.3.1   Procedure 1

Since $\bar{u} = -\Delta \bar{y}$, we simply take $\bar{u}_h$ to be $-\Delta_h \bar{y}_h$, where Δ_h is the piecewise Laplace operator with respect to $\mathcal{T}_h$.

2.3.2   Procedure 2

The optimal state $\bar{y}$ and the optimal control $\bar{u}$ are connected by

$$\int_\Omega \nabla \bar{y} \cdot \nabla z \, dx = \int_\Omega \bar{u} z \, dx \qquad \forall \, z \in H_0^1(\Omega)$$

for the Dirichlet problem and by

$$\int_\Omega \nabla \bar{y} \cdot \nabla z \, dx = \int_\Omega \bar{u} z \, dx \qquad \forall \, z \in H^1(\Omega)$$

for the Neumann problem. Therefore, we can compute an approximation $\bar{u}_h$ of $\bar{u}$ by solving

$$\int_\Omega \nabla \bar{y}_h \cdot \nabla z_h \, dx = \int_\Omega \bar{u}_h z_h \, dx \qquad \forall \, z_h \in \mathring{V}_h \tag{9}$$

for the Dirichlet problem and by solving

$$\int_\Omega \nabla \bar{y}_h \cdot \nabla z_h \, dx = \int_\Omega \bar{u}_h z_h \, dx \qquad \forall \, z_h \in V_h \tag{10}$$

for the Neumann problem.

2.3.3 Procedure 3

Here, we exploit the following relations between $\bar{y}$ and $\bar{u}$:

$$\int_\Omega \nabla \bar{u} \cdot \nabla z \, dx = -\int_\Omega \nabla(\Delta \bar{y}) \cdot \nabla z \, dx$$

$$= \int_\Omega (\Delta \bar{y})(\Delta z) \, dx$$

$$= \int_\Omega D^2 \bar{y} : D^2 z \, dx \quad \forall \, z \in V_D$$

for the Dirichlet problem and

$$\int_\Omega \nabla \bar{u} \cdot \nabla z \, dx = -\int_\Omega \nabla(\Delta \bar{y}) \cdot \nabla z \, dx$$

$$= \int_\Omega (\Delta \bar{y})(\Delta z) \, dx$$

$$= \int_\Omega D^2 \bar{y} : D^2 z \, dx \qquad \forall \, z \in V_N$$

for the Neumann problem. Therefore, we can compute an approximation $\bar{u}_h \in \mathring{V}_h$ of $\bar{u}$ by solving

$$\int_\Omega \nabla \bar{u}_h \cdot \nabla z_h \, dx = a_{D,h}(\bar{y}_h, z_h) \qquad \forall \, z_h \in \mathring{V}_h \tag{11}$$

for the Dirichlet problem and compute $\bar{u}_h \in V_h \cap L_0^2(\Omega)$ by solving

$$\int_\Omega \nabla \bar{u}_h \cdot \nabla z_h \, dx = a_{N,h}(\bar{y}_h, z_h) \qquad \forall \, z_h \in V_h \tag{12}$$

for the Neumann problem. Note that the solvability of (12) follows from the compatibility condition

$$a_{N,h}(\bar{y}_h, 1) = 0.$$

Remark 3 The computational cost increases from Procedure 1 to Procedure 3. However, these computational costs are negligible in comparison with the cost of solving the variational inequality.

3 Implementation

The discrete problems in the numerical experiments are solved by a primal-dual active set algorithm (cf. [18, 21] and the references therein).

Let $\bar{\mathbf{y}}_\star \in \mathbb{R}^N$ be the vector representing $\bar{y}_h$ in (5) (or (7)) with respect to a nodal basis of the $\mathbb{Q}_2$ finite element space, where N is the dimension of $\mathring{V}_h$ (or V_h). Similarly, $K \subset \mathbb{R}^N$ is the subset corresponding to $K_{D,h}$ (or $K_{N,h}$), and $\mathbf{A} \in \mathbb{R}^{N \times N}$ denotes the matrix representing the bilinear form $\mathcal{A}_{D,h}(\cdot, \cdot)$ (or $\mathcal{A}_{N,h}(\cdot, \cdot)$) with respect to the nodal basis of the $\mathbb{Q}_2$ finite element space. Then, (5) or (7) can be written as the following variational inequality: Find $\bar{\mathbf{y}}_\star \in K$ such that

$$(\mathbf{A}\bar{\mathbf{y}}_\star, \mathbf{y} - \bar{\mathbf{y}}_\star) \geq (\mathbf{f}, \mathbf{y} - \bar{\mathbf{y}}_\star) \qquad \forall \mathbf{y} \in K. \tag{13}$$

Here, $(\cdot, \cdot)$ is the Euclidean inner product on $\mathbb{R}^N$ and the vector $\mathbf{f}$ is defined by

$$(\mathbf{f}, \mathbf{y}) = \int_\Omega y_d y_h \, dx,$$

where the vector $\mathbf{y}$ represents the finite element function y_h in $\mathring{V}_h$ (or V_h).

Let $\boldsymbol{\lambda}_\star = \mathbf{f} - \mathbf{A}\bar{\mathbf{y}}_\star$. The primal-dual problem of (13) is to find $(\bar{\mathbf{y}}_\star, \boldsymbol{\lambda}_\star) \in \mathbb{R}^N \times \mathbb{R}^N$ such that

$$\begin{aligned}
\mathbf{A}\bar{\mathbf{y}}_\star + \boldsymbol{\lambda}_\star &= \mathbf{f}, \\
\boldsymbol{\psi} - \bar{\mathbf{y}}_\star &\geq \mathbf{0}, \\
\boldsymbol{\lambda}_\star &\geq \mathbf{0}, \\
(\boldsymbol{\lambda}_\star, \boldsymbol{\psi} - \bar{\mathbf{y}}_\star) &= \mathbf{0},
\end{aligned}$$

where $\boldsymbol{\psi}$ is a vector in $(\mathbb{R} \cup \{+\infty\})^N$ that represents the discrete constraint. In other words, the component of $\boldsymbol{\psi}$ corresponding to a node $p \in \mathcal{V}_h$ is given by $\psi(p)$, while all other components of $\boldsymbol{\psi}$ equal $+\infty$.

Equivalently, we can write

$$\begin{aligned}
\mathbf{A}\bar{\mathbf{y}}_\star + \boldsymbol{\lambda}_\star &= \mathbf{f}, \\
\bar{\mathbf{y}}_\star &= \boldsymbol{\psi} \quad \text{on } A_\star, \\
\boldsymbol{\lambda}_\star &= \mathbf{0} \quad \text{on } I_\star,
\end{aligned} \tag{14}$$

where A_* and I_* are the active set and inactive set defined, respectively, by

$$A_* = \left\{ j \in N : \lambda_*(j) = (\mathbf{f} - \mathbf{A}\bar{\mathbf{y}}_*)(j) > 0 \text{ and } \bar{\mathbf{y}}_*(j) = \boldsymbol{\psi}(j) \right\},$$
$$I_* = \left\{ j \in N : \lambda_*(j) = (\mathbf{f} - \mathbf{A}\bar{\mathbf{y}}_*)(j) = 0 \text{ and } \bar{\mathbf{y}}_*(j) \le \boldsymbol{\psi}(j) \right\}.$$

Here, $N = \{1, 2, \dots, N\}$ and $\lambda_*(j)$ is the jth component of λ_*.

The primal-dual active set method solves (14) by generating a sequence of sets A_k and I_k that approximate A_* and I_* and then obtain the approximation $(\bar{\mathbf{y}}_k, \lambda_k)$ by solving a reduced system.

Given an initial guess $(\bar{\mathbf{y}}_0, \lambda_0) \in \mathbb{R}^N \times \mathbb{R}^N$ where $\lambda_0 \ge 0$, we define

$$A_0 = \left\{ j \in N : \lambda_0(j) + c(\bar{\mathbf{y}}_0(j) - \boldsymbol{\psi}(j)) > 0 \right\},$$
$$I_0 = \left\{ j \in N : \lambda_0(j) + c(\bar{\mathbf{y}}_0(j) - \boldsymbol{\psi}(j)) \le 0 \right\} = N \setminus A_0,$$

where c is a positive number.

For $k \ge 1$, we solve the reduced system

$$\begin{aligned}
\mathbf{A}_k \bar{\mathbf{y}}_k + \lambda_k &= \mathbf{f}_k, \\
\bar{\mathbf{y}}_k &= \boldsymbol{\psi} \quad \text{on } A_{k-1}, \\
\lambda_k &= \mathbf{0} \quad \text{on } I_{k-1},
\end{aligned} \qquad (15)$$

and update the active set and inactive set by

$$A_k = \left\{ j \in N : \lambda_k(j) + c(\bar{\mathbf{y}}_k(j) - \boldsymbol{\psi}(j)) > 0 \right\}, \qquad (16)$$
$$I_k = \left\{ j \in N : \lambda_k(j) + c(\bar{\mathbf{y}}_k(j) - \boldsymbol{\psi}(j)) \le 0 \right\} = N \setminus A_k. \qquad (17)$$

(Choosing $c > 0$ large, e.g., $c = 10^7$, can improve the performance of the computation.)

Let the diagonal matrices $\mathbf{P}_{A_k}, \mathbf{P}_{I_k} \in \mathbb{R}^{N \times N}$ be defined by

$$[\mathbf{P}_{A_k} \mathbf{v}](j) = \begin{cases} \mathbf{v}(j) & \text{if } j \in A_k, \\ 0 & \text{if } j \in I_k, \end{cases}$$
$$[\mathbf{P}_{I_k} \mathbf{v}](j) = \begin{cases} \mathbf{v}(j) & \text{if } j \in I_k, \\ 0 & \text{if } j \in A_k. \end{cases}$$

Then solving (15) is equivalent to solving

$$\mathbf{P}_{I_{k-1}} \mathbf{A}_k \mathbf{P}_{I_{k-1}} (\mathbf{P}_{I_{k-1}} \bar{\mathbf{y}}_k) = \mathbf{P}_{I_{k-1}} \mathbf{f}_k - \mathbf{P}_{I_{k-1}} \mathbf{A}_k \mathbf{P}_{A_{k-1}} \boldsymbol{\psi} \qquad (18)$$

together with

$$\mathbf{P}_{A_{k-1}} \bar{\mathbf{y}}_k = \mathbf{P}_{A_{k-1}} \boldsymbol{\psi}, \quad \mathbf{P}_{I_{k-1}} \lambda_k = \mathbf{0} \quad \text{and} \quad \mathbf{P}_{A_{k-1}} \lambda_k = \mathbf{P}_{A_{k-1}} \mathbf{f}_k - \mathbf{P}_{A_{k-1}} \mathbf{A}_k \bar{\mathbf{y}}_k.$$

The iteration is terminated when two consecutive active sets determined by (16) are identical. The linear system (18) is solved by the preconditioned conjugate gradient method with an algebraic multigrid preconditioner, implemented within the Trilinos library [17].

For the problem on the coarsest mesh $\mathcal{T}_0$, all degrees of freedom are initially placed in the inactive set. For subsequent refinements at level $k \geq 1$, we first compute $\tilde{\mathbf{y}}_k$ and $\tilde{\boldsymbol{\lambda}}_k$ from $\mathbf{y}_{k-1}$ and $\boldsymbol{\lambda}_{k-1}$ through interpolation, and then, we initialize the active and inactive set using (16) and (17), with $\mathbf{y}_k$ and $\boldsymbol{\lambda}_k$ replaced by $\tilde{\mathbf{y}}_k$ and $\tilde{\boldsymbol{\lambda}}_k$, respectively.

To speed up the solution of the linear system, an inexact method is implemented in which the inner iteration runs to a tolerance determined by the maximum of an absolute tolerance and a relative tolerance based on the norm of the initial residual from (15). This approach was observed to yield a solution in fewer iterations than either solving to a uniform absolute tolerance alone or to a relative tolerance based on the residual of the inner iteration alone. With this approach, the solver required less than 8 iterations of the primal-dual active set method for the examples presented in Section 4.

The numerical implementation has been realized by using the C++ software library deal.II [1, 2]. The skeleton of the code is based on the deal.II tutorial step-41, while the assembling of the local cell and face matrices relies on the `LocalIntegrators` classes within the `MeshWorker` framework (formally introduced in tutorial step-39).

Since the assembled matrices correspond to the C^0 interior penalty formulation of the biharmonic operator, which are different from the one implemented in deal.II, we rely on the `LocalIntegrators` for the weak form of this problem. Furthermore, the calculation of higher order derivatives is performed by using the `contract` family of deal.II functions.

4   Numerical Results

In this section, we present numerical examples for (3). The discrete optimal state $\bar{y}_h$ is obtained from (5) for the Dirichlet problem and (7) for the Neumann problem, and we use the post-processing procedures in Section 2.3 to generate the discrete optimal control $\bar{u}_h$. For each example, we report the state error in a H^2-like mesh-dependent norm (cf. (19) and (20)) and in the H^1, L^2, and L^∞-norms. We also report the L^2 control errors for all post-processing procedures, and the H^1 control errors of Procedures 2 and 3. Finally, we present a figure of the contact set for each example. We will comment on the numerical results in Section 5.

Examples 1–3 correspond to the optimal control problem with the Dirichlet boundary condition, and Examples 4–6 are concerned with the Neumann boundary condition. The domain is the unit cube $\Omega = (-0.5, 0.5)^3$ for all the examples.

We will use $\|\cdot\|_\infty$ to denote the ℓ^∞ norm defined by

$$\|v\|_\infty = \max_{p \in \mathcal{N}_h} |v(p)|,$$

where $\mathcal{N}_h$ is the set of the nodes of the $\mathbb{Q}_2$ finite element space associated with $\mathcal{T}_h$, and we define the mesh-dependent norms $\|\cdot\|_{h,D}$ and $\|\cdot\|_{h,N}$ by

$$\|v\|_{h,D}^2 = \sum_{T \in \mathcal{T}_h} \|D^2 v\|_{L_2(T)}^2 + \sum_{F \in \mathcal{F}_h} \frac{1}{h_T} \left\| \left[\!\!\left[\frac{\partial v}{\partial n} \right]\!\!\right] \right\|_{L_2(F)}^2, \tag{19}$$

$$\|v\|_{h,N}^2 = \sum_{T \in \mathcal{T}_h} \|D^2 v\|_{L_2(T)}^2 + \sum_{F \in \mathcal{F}_h} \frac{1}{h_T} \left\| \left[\!\!\left[\frac{\partial v}{\partial n} \right]\!\!\right] \right\|_{L_2(F)}^2. \tag{20}$$

We solve the discrete problems on a sequence of triangulations generated by uniform refinements, where the coarsest mesh consists of a single element. The number of degrees of freedom at the kth level is $(2k + 3)^3$. The discrete optimal state associated with $\mathcal{T}_k$ is denoted by $\bar{y}_k$, the discrete optimal control obtained by the post-processing procedure i ($1 \leq i \leq 3$) is denoted by $\bar{u}_{k,i}$, and N_k stands for the number of degrees of freedom at mesh level k. The word "order" denotes the order of convergence computed by $\ln(\|e_{k-1}\| / \|e_k\|)/\ln 2$, where $e_k = \bar{y} - \bar{y}_k$ (or $e_k = \bar{u} - \bar{u}_{k,j}$) if the exact optimal state $\bar{y}$ (or the exact optimal control $\bar{u}$) is available. If the exact solution is not available, then we take $e_k = \bar{y}_k - \bar{y}_{k-1}$ (or $e_k = \bar{u}_{k,i} - \bar{u}_{k-1,i}$).

The CPU time for the 3D computations shown was observed to increase linearly with the number of degrees of freedom. For each example, the numerical results on the finest-level mesh took approximately 18 hours to complete. All the results below were generated on the SuperMIC at Louisiana State University without using parallel processing.

Example 1 (Dirichlet problem with a known solution)

We begin by considering (3) on the ball of radius two centered at the origin. We take γ to be 1 and the exact solution to be

$$\bar{y} = \begin{cases} r^2 - 1 & \text{if } r \leq r_0 \\ \dfrac{1}{120} r^4 + C_1 + C_2 r + C_3 r^2 + C_4/r & \text{if } r > r_0 \end{cases},$$

where $r = \sqrt{x_1^2 + x_2^2 + x_3^2}$, $r_0 = 0.32151559$, $C_1 = -1.4090715$, $C_2 = 1.2737074$, $C_3 = -0.32339567$, and $C_4 = 0.043812326$. The upper bound for the state is given by $\psi = r^2 - 1$, and the desired state y_d is $1 + \bar{y}$.

The restriction of $\bar{y}$ to the unit cube $\Omega = (-0.5, 0.5)^3$ is the exact solution of (3) with the same y_d and ψ, but the nonhomogeneous boundary conditions determined by $\bar{y}$.

Remark 4 The exact solution (3) on the ball is obtained by reducing the problem to a one-dimensional problem through the rotational symmetry.

Table 1 State errors for Example 1.

k	N_k	$\|\bar{y} - \bar{y}_k\|_{h,D}$	order	$\|\bar{y} - \bar{y}_k\|_{H^1(\Omega)}$	order	$\|\bar{y} - \bar{y}_k\|$	order	$\|\bar{y} - \bar{y}_k\|_\infty$	order
1	27	2.557e+00	–	3.148e−01	–	4.770e−02	–	2.433e−01	–
2	125	2.667e+00	−0.062	1.692e−01	0.896	2.897e−02	0.720	6.565e−02	1.890
3	729	1.061e+00	1.330	2.571e−02	2.718	2.980e−03	3.281	6.867e−03	3.257
4	4913	4.337e−01	1.290	7.466e−03	1.784	5.068e−04	2.556	2.155e−03	1.672
5	35940	1.800e−01	1.268	1.526e−03	2.290	7.446e−05	2.767	4.309e−04	2.322
6	274625	7.988e−02	1.172	3.722e−04	2.036	1.915e−05	1.959	8.420e−05	2.355
7	2146689	3.610e−02	1.146	8.528e−05	2.126	3.033e−06	2.658	1.250e−05	2.752

Table 2 L^2 control errors for the three procedures, Example 1.

k	N_k	$\|\bar{u} - \bar{u}_{k,1}\|$	order	$\|\bar{u} - \bar{u}_{k,2}\|$	order	$\|\bar{u} - \bar{u}_{k,3}\|$	order
1	27	1.290e+00	–	1.240e+00	–	1.260e+00	–
2	125	1.078e+00	0.259	1.249e+00	−0.109	1.354e+00	−0.104
3	729	4.958e−01	1.120	3.974e−01	1.652	4.473e−01	1.598
4	4913	2.512e−01	0.981	1.600e−01	1.313	1.726e−01	1.374
5	35940	1.227e−01	1.033	5.435e−02	1.558	5.717e−02	1.594
6	274625	6.139e−02	0.999	2.020e−02	1.428	2.112e−02	1.437
7	2146689	3.065e−02	1.002	7.617e−03	1.407	7.480e−03	1.498

In view of the nonhomogeneous boundary conditions, we change the definition of $K_{D,h}$ to

$$K_{D,h} = \{y \in V_h : y - \Pi_h \bar{y} \in H_0^1(\Omega) \text{ and } y(p) \le \psi(p) \text{ for all } p \in \mathcal{V}_h\},$$

where Π_h is the Lagrange nodal interpolation operator. The discrete problem (5) then becomes

$$\text{Find } \bar{y}_h = \operatorname*{argmin}_{y_h \in K_{D,h}} \left[\frac{1}{2} \mathcal{A}_{N,h}(y_h, y_h) - \int_{\partial\Omega} \frac{\partial^2 \bar{y}}{\partial n^2} \frac{\partial y_h}{\partial n} \, dS - (f, y_h) \right].$$

In Table 1, we report the error of the state in $\| \cdot \|_{h,D}$ and in the H^1, L^2, and L^∞-norms. In Table 2, we report L^2 control errors for all post-processing procedures described in Section 2.3, and we report the H^1 control errors of Procedures 2 and 3 in Table 3. The discrete contact set after 4 uniform refinements is shown in Figure 1.

Example 2 (Dirichlet problem with an unknown solution)

We take γ to be 10^{-3}, ψ to be the constant 0.2, and y_d to be the function $\sin(2\pi(x_1 + 0.5)(x_2 + 0.5)(x_3 + 0.5))$. The errors for the state and the control are reported in Tables 4–6. The discrete contact set after 4 uniform refinements is displayed in Figure 2.

Table 3 H^1 control errors for Procedures 2 and 3, Example 1.

k	N_k	$\|\bar{u} - \bar{u}_{k,2}\|_{H^1(\Omega)}$	order	$\|\bar{u} - \bar{u}_{k,3}\|_{H^1(\Omega)}$	order
1	27	8.961e+00	–	9.055e+00	–
2	125	1.573e+01	−0.812	1.852e+01	−0.103
3	729	1.444e+01	0.123	1.601e+01	0.210
4	4913	1.043e+01	0.469	1.129e+01	0.504
5	35940	7.438e+00	0.488	8.021e+00	0.493
6	274625	5.501e+00	0.435	5.774e+00	0.474
7	2146689	4.212e+00	0.385	4.039e+00	0.516

Fig. 1 The discrete contact set for Example 1 after 4 uniform refinements.

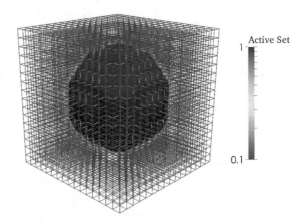

Active Set

Example 3 (Dirichlet problem with an unknown solution)

As in Example 2, we take γ to be 10^{-3} and the upper bound ψ to be a constant 0.15. But we choose y_d to be a piecewise constant function:

$$y_d = \begin{cases} 0 & \text{if } x < 0, \\ 1 & \text{otherwise.} \end{cases}$$

The errors for the state and the control are reported in Tables 7–9. The discrete contact set after 4 levels of uniform refinement is shown in Figure 3.

Example 4 (Neumann problem with an exact solution)

As in Example 1, we begin with (3) on the ball of radius 2 centered at the origin. We take γ to be 1 and the exact solution to be

$$\bar{y} = \begin{cases} C_1 r + C_2 r^2 + C_3/r + C_4 & \text{if } r > r_0 \\ r^2 - r^4/8 & \text{if } r \le r_0 \end{cases},$$

Table 4 State errors for Example 2.

k	N_k	$\|y_{k-1} - y_k\|_{h,D}$	order	$\|y_{k-1} - y_k\|_{H^1(\Omega)}$	order	$\|y_{k-1} - y_k\|$	order	$\|y_{k-1} - y_k\|_\infty$	order
2	125	6.033e+00	–	4.442e−01	–	6.535e−02	–	3.405e−01	–
3	729	4.798e+00	0.330	2.948e−01	0.591	4.473e−02	0.547	1.654e−01	1.041
4	4913	4.044e+00	0.247	1.237e−01	1.253	1.684e−02	1.410	7.936e−02	1.060
5	35940	1.640e+00	1.302	2.556e−02	2.275	1.527e−03	3.463	1.015e−02	2.967
6	274625	7.264e−01	1.175	5.580e−03	2.196	2.549e−04	2.583	1.645e−03	2.625
7	2146689	3.004e−01	1.274	1.521e−03	1.876	7.2393e−05	1.816	2.521e−04	2.706

Table 5 L^2 control errors for the three procedures, Example 2.

k	N_k	$\|\bar{u}_{k-1,1} - \bar{u}_{k,1}\|$	order	$\|\bar{u}_{k-1,2} - \bar{u}_{k,2}\|$	order	$\|\bar{u}_{k-1,3} - \bar{u}_{k,3}\|$	order
2	125	1.993e+00	–	4.080e+00	–	4.657e+00	–
3	729	1.531e+00	0.380	3.080e+00	0.406	3.418e+00	0.446
4	4913	1.285e+00	0.253	2.086e+00	0.563	2.341e+00	0.546
5	35940	7.869e−01	0.707	7.617e−01	1.453	8.240e−01	1.506
6	274625	4.135e−01	0.928	2.501e−01	1.607	2.752e−01	1.582
7	2146689	2.108e−01	0.972	7.933e−02	1.657	8.105e−02	1.763

Table 6 H^1 control errors for Procedures 2 and 3, Example 2.

k	N_k	$\|\bar{u}_{k-1,2} - \bar{u}_{k,2}\|_{H^1(\Omega)}$	order	$\|\bar{u}_{k-1,3} - \bar{u}_{k,3}\|_{H^1(\Omega)}$	order
2	125	4.763e+01	–	5.486e+01	–
3	729	6.914e+01	−0.538	7.485e+01	−0.448
4	4913	7.972e+01	−0.205	8.584e+01	−0.198
5	35940	5.986e+01	0.413	6.342e+01	0.437
6	274625	3.941e+01	0.603	4.107e+01	0.627
7	2146689	2.537e+01	0.635	2.610e+01	0.654

where $r_0 = 0.33563105$, $C_1 = 1.2785390$, $C_2 = -0.31672296$, $C_3 = 0.046588697$, and $C_4 = -0.42118638$. The upper bound for the state is given by $\psi = r^2 - r^4/8$, and the desired state y_d equals $\bar{y}$.

The restriction of $\bar{y}$ to the unit cube $\Omega = (-0.5, 0.5)^3$ is the exact solution of (3) with the same ψ and y_d, and the nonhomogeneous boundary conditions determined by $\bar{y}$.

In view of the nonhomogeneous conditions, the discrete problem (7) becomes

$$\text{Find } \bar{y}_h = \underset{y_h \in K_{N,h}}{\operatorname{argmin}} \left[\frac{1}{2} \mathcal{A}_{N,h}(y_h, y_h) + \int_{\partial\Omega} \frac{\partial \Delta \bar{y}}{\partial n} y_h \, dS \right.$$

$$\left. - \int_{\partial\Omega} \left(\frac{\partial}{\partial n} \nabla_{\partial\Omega} \bar{y} \right) \cdot \nabla_{\partial\Omega} y_h \, dS - (f, y_h) \right].$$

Fig. 2 The discrete contact set for Example 2 after 4 uniform refinements.

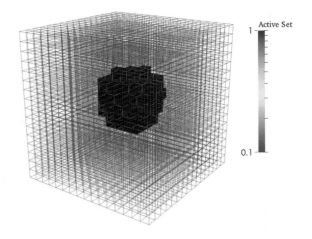

Table 7 State errors for Example 3.

k	N_k	$\|y_{k-1} - y_k\|_{h,D}$	order	$|y_{k-1} - y_k|_{H^1(\Omega)}$	order	$\|y_{k-1} - y_k\|$	order	$|y_{k-1} - y_k|_\infty$	order
2	125	9.992e+00	–	1.145e+00	–	1.937e−01	–	6.804e−01	–
3	729	6.583e+00	0.602	4.788e−01	1.257	7.286e−02	1.410	2.999e−01	1.182
4	4913	4.393e+00	0.583	1.358e−01	1.819	1.732e−02	2.072	7.681e−02	1.965
5	35937	1.934e+00	1.184	3.130e−02	2.117	1.777e−03	3.285	1.049e−02	2.872
6	274625	8.585e−01	1.171	8.125e−03	1.946	4.845e−04	1.875	1.455e−03	2.850
7	2146689	3.735e−01	1.201	1.892e−03	2.102	1.030e−04	2.234	3.053e−04	2.252

Table 8 L^2 control errors for the three procedures, Example 3.

k	N_k	$\|\bar{u}_{k-1,1} - \bar{u}_{k,1}\|$	order	$\|\bar{u}_{k-1,2} - \bar{u}_{k,2}\|$	order	$\|\bar{u}_{k-1,3} - \bar{u}_{k,3}\|$	order
2	125	4.425e+00	–	7.644e+00	–	8.513e+00	–
3	729	2.382e+00	0.893	4.359e+00	0.810	4.823e+00	0.820
4	4913	1.280e+00	0.896	2.417e+00	0.851	2.663e+00	0.857
5	35940	9.293e−01	0.462	9.022e−01	1.422	8.879e−01	1.585
6	274625	5.111e−01	0.863	3.045e−01	1.567	3.013e−01	1.560
7	2146689	2.630e−01	0.958	1.006e−01	1.597	9.998e−02	1.591

Table 9 H^1 control errors for Procedures 2 and 3, Example 3.

k	N_k	$|\bar{u}_{k-1,2} - \bar{u}_{k,2}|_{H^1(\Omega)}$	order	$|\bar{u}_{k-1,3} - \bar{u}_{k,3}|_{H^1(\Omega)}$	order
2	125	6.299e+01	–	7.138e+01	–
3	729	8.628e+01	−0.454	9.310e+01	−0.383
4	4913	9.490e+01	−0.137	1.006e+02	−0.112
5	35940	7.016e+01	0.436	7.221e+01	0.479
6	274625	4.786e+01	0.552	4.919e+01	0.554
7	2146689	3.171e+01	0.594	3.223e+01	0.610

Fig. 3 The discrete contact
set for Example 3 after 4
uniform refinements.

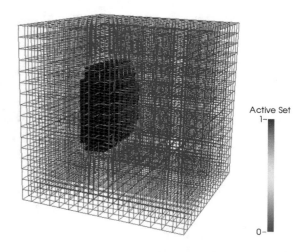

Active Set
1—

0—

Table 10 State errors for Example 4.

k	N_k	$\|\bar{y} - \bar{y}_k\|_{h,N}$	order	$\|\bar{y} - \bar{y}_k\|_{H^1(\Omega)}$	order	$\|\bar{y} - \bar{y}_k\|$	order	$\|\bar{y} - \bar{y}_k\|_\infty$	order
1	27	8.161e+00	–	4.445e−01	–	6.470e−01	–	7.204e−01	–
2	125	3.439e+00	1.247	9.993e−02	2.153	5.137e−02	3.655	7.112e−02	3.341
3	729	1.208e+00	1.509	2.954e−02	1.758	3.582e−03	3.842	7.202e−03	3.304
4	4913	4.315e−01	1.485	8.196e−03	1.850	9.221e−04	1.958	2.786e−03	1.370
5	35940	1.780e−01	1.278	1.774e−03	2.208	2.323e−04	1.989	7.165e−04	1.959
6	274625	7.654e−02	1.217	4.465e−04	1.990	6.995e−05	1.732	2.030e−04	1.819
7	2146689	3.480e−02	1.137	1.084e−04	2.043	1.790e−05	1.967	5.138e−05	1.982

Table 11 L^2 control errors for the three procedures, Example 4.

k	N_k	$\|\bar{u} - \bar{u}_{k,1}\|$	order	$\|\bar{u} - \bar{u}_{k,2}\|$	order	$\|\bar{u} - \bar{u}_{k,3}\|$	order
1	27	1.548e+00	–	4.019e+00	–	5.234e+01	–
2	125	9.684e−01	0.676	1.389e+00	1.533	2.418e+00	4.436
3	729	4.933e−01	0.973	4.373e−01	1.667	4.476e−01	2.433
4	4913	2.411e−01	1.033	1.582e−01	1.467	1.613e−01	1.473
5	35940	1.194e−01	1.014	5.288e−02	1.581	5.512e−02	1.549
6	274625	5.938e−02	1.008	1.864e−02	1.504	1.927e−02	1.516
7	2146689	2.967e−02	1.001	7.141e−03	1.384	6.755e−03	1.512

The errors for the state are summarized in Table 10, while Tables 11 and 12
contain the L^2 and H^1 errors for the control for the post-processing procedures
from Section 2.3. The discrete contact set after 4 uniform refinements is depicted in
Figure 4.

Table 12 H^1 control errors for Procedures 2 and 3, Example 4.

k	N_k	$|\bar{u} - \bar{u}_{k,2}|_{H^1(\Omega)}$	order	$|\bar{u} - \bar{u}_{k,3}|_{H^1(\Omega)}$	order
1	27	4.143e+01	–	4.203e+01	–
2	125	2.708e+01	0.614	2.743e+01	0.615
3	729	1.560e+01	0.795	1.628e+01	0.753
4	4913	1.025e+01	0.607	1.078e+01	0.595
5	35940	7.160e+00	0.517	7.601e+00	0.504
6	274625	4.944e+00	0.534	5.222e+00	0.542
7	2146689	4.034e+00	0.294	3.671e+00	0.509

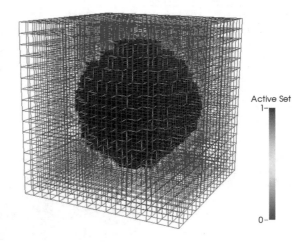

Fig. 4 The discrete contact set for Example 4 after 4 levels of uniform refinement.

Example 5 (Neumann problem with an unknown solution)

We take γ to be 10^{-3}, the upper bound ψ to be the constant 0.2, and y_d to be the function $\sin(2\pi(x_1 + 0.5)(x_2 + 0.5)(x_3 + 0.5))$. The errors for the state and the control are reported in Tables 13–15. The discrete contact set after 4 uniform refinements is shown in Figure 5. Note that the contact set is not disjoint from $\partial\Omega$ for this example.

Table 13 State errors for Example 5.

k	N_k	$\|y_{k-1} - y_k\|_{h,N}$	order	$|y_{k-1} - y_k|_{H^1(\Omega)}$	order	$|y_{k-1} - y_k|$	order	$|y_{k-1} - y_k|_\infty$	order
2	125	1.346e+01	–	5.244e−01	–	1.327e−01	–	3.772e−01	–
3	729	1.137e+01	0.244	3.885e−01	0.433	1.231e−01	0.109	2.634e−01	0.518
4	4913	4.886e+00	1.218	1.157e−01	1.747	1.807e−02	2.768	4.535e−02	2.538
5	35940	1.469e+00	1.734	1.557e−02	2.894	1.403e−03	3.687	4.255e−03	3.414
6	274625	4.959e−01	1.566	2.890e−03	2.430	2.106e−04	2.736	1.610e−03	1.402
7	2146689	1.801e−01	1.461	6.627e−04	2.125	5.012e−05	2.071	4.322e−04	1.897

Table 14 L^2 control errors for the three procedures, Example 5.

k	N_k	$\|\bar{u}_{k-1,1} - \bar{u}_{k,1}\|$	order	$\|\bar{u}_{k-1,2} - \bar{u}_{k,2}\|$	order	$\|\bar{u}_{k-1,3} - \bar{u}_{k,3}\|$	order
2	125	2.427e+00	–	5.291e+00	–	5.448e+00	–
3	729	2.496e+00	−0.046	5.194e+00	0.027	8.394e+00	−0.624
4	4913	1.341e+00	0.897	2.671e+00	0.960	3.297e+00	1.348
5	35940	4.637e−01	1.532	7.730e−01	1.789	9.107e−01	1.856
6	274625	2.102e−01	1.141	2.165e−01	1.836	2.593e−01	1.812
7	2146689	1.027e−01	1.034	6.253e−02	1.792	7.846e−02	1.724

Table 15 H^1 control errors for Procedures 2 and 3, Example 5.

k	N_k	$|\bar{u}_{k-1,2} - \bar{u}_{k,2}|_{H^1(\Omega)}$	order	$|\bar{u}_{k-1,3} - \bar{u}_{k,3}|_{H^1(\Omega)}$	order
2	125	9.239e+01	–	9.350e+01	–
3	729	1.291e+02	−0.483	1.327e+02	−0.505
4	4913	1.127e+02	0.197	1.155e+02	0.201
5	35940	6.735e+01	0.743	6.879e+01	0.747
6	274625	3.771e+01	0.837	3.870e+01	0.830
7	2146689	2.177e+01	0.793	2.224e+01	0.799

Fig. 5 The discrete contact set for Example 5 after 4 uniform refinements.

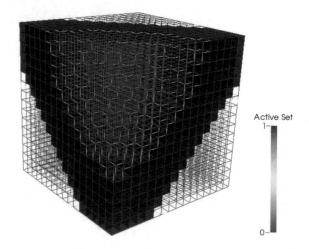

Example 6 (Neumann problem with an unknown solution)

For the last example, we take γ to be 10^{-1} and the upper bound ψ to be the constant 0.5. Unlike Example 5, we choose y_d to be the piecewise constant function defined by

$$y_d = \begin{cases} 20 & \text{if } (x, y, z) \in (-0.375, 0.125)^3, \\ 0 & \text{otherwise.} \end{cases}$$

Table 16 State errors for Example 6.

k	N_k	$\|y_{k-1}-y_k\|_{h,N}$	order	$\|y_{k-1}-y_k\|_{H^1(\Omega)}$	order	$\|y_{k-1}-y_k\|$	order	$\|y_{k-1}-y_k\|_\infty$	order
2	125	3.833e+00	–	1.735e−01	–	1.574e−01	–	2.048e−01	–
3	729	1.959e+00	0.968	5.836e−02	1.572	3.642e−02	2.112	6.586e−02	1.637
4	4913	5.081e−01	1.947	3.812e−02	0.615	1.584e−02	1.201	2.659e−02	1.309
5	35940	2.122e−01	1.260	2.669e−03	3.836	6.442e−04	4.620	1.035e−03	4.684
6	274625	7.672e−02	1.468	5.823e−04	2.196	1.401e−04	2.201	3.610e−04	1.519
7	2146689	2.961e−02	1.374	1.329e−04	2.131	3.205e−05	2.128	7.000e−05	2.367

Table 17 L^2 control errors for the three procedures, Example 6.

k	N_k	$\|\bar{u}_{k-1,1}-\bar{u}_{k,1}\|$	order	$\|\bar{u}_{k-1,2}-\bar{u}_{k,2}\|$	order	$\|\bar{u}_{k-1,3}-\bar{u}_{k,3}\|$	order
2	125	6.761e−01	–	1.638e+00	–	3.688e+00	–
3	729	4.424e−01	0.612	8.761e−01	0.903	1.508e+00	1.290
4	4913	2.646e−01	0.741	3.320e−01	1.400	3.789e−01	1.993
5	35940	6.588e−02	2.006	1.157e−01	1.521	1.245e−01	1.605
6	274625	2.980e−02	1.144	3.538e−02	1.709	3.951e−02	1.656
7	2146689	1.490e−02	1.000	1.071e−02	1.724	1.217e−02	1.699

Table 18 H^1 control errors for Procedures 2 and 3, Example 6.

k	N_k	$\|\bar{u}_{k-1,2}-\bar{u}_{k,2}\|_{H^1(\Omega)}$	order	$\|\bar{u}_{k-1,3}-\bar{u}_{k,3}\|_{H^1(\Omega)}$	order
2	125	2.132e+01	–	2.145e+01	–
3	729	1.809e+01	0.237	1.879e+01	0.191
4	4913	1.052e+01	0.783	1.102e+01	0.770
5	35940	9.810e+00	0.100	1.005e+01	0.133
6	274625	5.928e+00	0.727	6.165e+00	0.705
7	2146689	3.620e+00	0.712	3.687e+00	0.742

We report the errors for the state and the control in Tables 16–18. The discrete contact set after 4 uniform refinements is displayed in Figure 6. Note that the contact set is not disjoint from $\partial\Omega$ for this example.

Fig. 6 The discrete contact set for Example 6 after 4 uniform refinements.

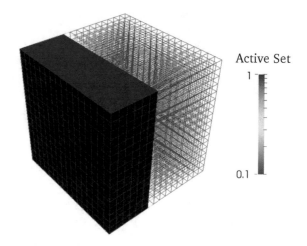

Active Set

1

0.1

5 Concluding Remarks

We have obtained the first numerical results for solving elliptic distributed optimal control problems in three-dimensional domains as fourth order variational inequalities.

In all the examples, we have observed $O(h)$ convergence for the state in the mesh-dependent norm $\| \cdot \|_{h,D}$ or $\| \cdot \|_{h,N}$. This is most evident for Example 1 and Example 4 where we know the exact solutions. The convergence in the other examples appears to be slightly better than $O(h)$, probably because the errors in these examples are only estimated by comparing the solutions on consecutive levels.

Note that the exact solution $\bar{y}$ belongs to $H^3_{loc}(\Omega)$ by the result in [12] for fourth order variational inequalities. When the contact set is disjoint from the boundary of the unit cube (Examples 1–4), $\bar{y}$ belongs globally to $H^3(\Omega)$ by the result in [10, 25] for elliptic boundary value problems on nonsmooth domains. Therefore, an $O(h)$ error in the H^2-like mesh-dependent norm for a method based on the $\mathbb{Q}_2$ element is not surprising. On the other hand, when the contact set is not disjoint from the boundary of the domain, the global regularity of the exact solution $\bar{y}$ is very much problem dependent. The observed convergence behavior for Examples 5 and 6 indicates that the exact solution of the two optimal control problems in these examples may also belong to $H^3(\Omega)$.

The convergence for the state in the lower order norms is of higher order in all the examples. In particular, the H^1 error of the state is $O(h^2)$ in all the examples, which compares favorably to the $O(h)$ error in standard finite element methods for optimal control problems where the state y is eliminated (cf. [19] and the references therein).

For the approximate control generated by post-processing, the convergence in the L^2 norm is $O(h)$ for Procedure 1 and $O(h^{3/2})$ for Procedure 2 and Procedure 3. We also observe that up to two hundred thousand degrees of freedom, the magnitudes of the L^2 errors for the approximate control generated by Procedure 1 in Experiment 5 and Experiment 6 are smaller than those for the other two procedures.

As in the two dimensional case, we also observe that the approximate optimal controls generated by Procedure 2 and Procedure 3 converge in the H^1 norm. This phenomenon has not been observed in the standard approach.

Finally, we remark that a direct extension of the methodology developed in [9] only leads to $O(h^{\frac{1}{2}})$ convergence in the mesh-dependent energy norms when the exact solution $\bar{y}$ belongs to $H^3(\Omega)$. New techniques are required for the analysis of three-dimensional problems.

Acknowledgments The work of the first author was supported in part by the National Science Foundation under Grant No. DMS-13-19172.

References

1. Bangerth, W., Hartmann, R., Kanschat, G., deal.II – a General Purpose Object Oriented Finite Element Library. ACM Trans. Math. Softw., **33**, 24/1–24/27, (2007)
2. Bangerth, W., Heister, T., Heltai, L., Kanschat, G., Kronbichler, M., Maier, M., Turcksin, B., Young, T. D., The deal.II Library, version 8.2. Archive of Numerical Software, **3**, (2015)
3. Brenner, S. C., C^0 Interior Penalty Methods, Frontiers in Numerical Analysis-Durham 2010, Springer-Verlag, Berlin-Heidelberg, **85**, 79–147, (2012)
4. Brenner, S. C., Gu, S., Gudi, T., Sung, L.-Y., A quadratic C^0 interior penalty method for linear fourth order boundary value problems with boundary conditions of the Cahn-Hilliard type. SIAM J. Numer. Anal., **50**, 2088–2110, (2012)
5. Brenner, S. C., Neilan, M., A C^0 interior penalty method for a fourth order elliptic singular perturbation problem. SIAM J. Numer. Anal., **49**, 869–892, (2011)
6. Brenner, S. C., Sung, L.-Y., C^0 interior penalty methods for fourth order elliptic boundary value problems on polygonal domains. J. Sci. Comput. **22/23**, 83–118, (2005)
7. Brenner, S. C., Sung, L.-Y., Zhang, H., Zhang, Y., A quadratic C^0 interior penalty method for the displacement obstacle problem of clamped Kirchhoff plates. SIAM J. Numer. Anal., **50**, 3329–3350, (2012)
8. Brenner, S. C., Sung, L.-Y., Zhang, Y., A quadratic C^0 interior penalty method for an elliptic optimal control problem with state constraints. Recent Developments in Discontinuous Galerkin Finite Element Methods for Partial Differential Equations, Feng, X., Karakashian, O. and Xing, Y. eds.,The IMA Volumes in Mathematics and its Applications, Springer International Publishing, **157**, 97–132, (2014)
9. Brenner, S. C., Sung, L.-Y., Zhang, Y., Post-processing procedures for an elliptic distributed optimal control problem with pointwise state constraints. Appl. Numer. Math., **95**, 99–117, (2015)
10. Dauge, M., Elliptic Boundary Value Problems on Corner Domains, Lecture Notes in Mathematics 1341, Springer-Verlag, Berlin-Heidelberg, (1988)
11. Engel, G., Garikipati, K., Hughes, T. J. R., Larson, M. G., Mazzei, L., Taylor, R. L., Continuous/discontinuous finite element approximations of fourth order elliptic problems in structural and continuum mechanics with applications to thin beams and plates, and strain gradient elasticity. Comput. Methods Appl. Mech. Engrg., **191**, 3669–3750, (2002)
12. Frehse, J., Zum Differenzierbarkeitsproblem bei Variationsungleichungen höherer Ordnung. Abh. Math. Sem. Univ. Hamburg, **36**, 140–149, (1971)
13. Friedman, A., Variational Principles and Free-Boundary Problems. Robert E. Krieger Publishing Co., Inc., Malabar, FL, second edition, (1988)

14. Gong, W., Yan, N., A mixed finite element scheme for optimal control problems with pointwise state constraints. J. Sci. Comput., **46**, 182–203, (2011)
15. Grisvard, P., Elliptic Problems in Non Smooth Domains. Pitman, Boston (1985)
16. Gudi, T., Gupta, H., Nataraj, N., Analysis of an interior penalty method for fourth order problems on polygonal domains. J. Sci. Comp. **54**, 177–199 (2013)
17. A. Heroux, M. A., Willenbring, J. M., Trilinos Users Guide, Sandia National Laboratories, (2003)
18. Hintermüller, M., Ito, K., Kunisch, K., The primal-dual active set strategy as a semismooth Newton method., SIAM J. Optim., **13**, 865–888, (2003)
19. Hinze, M., Pinnau, R.,Ulbrich, M., Ulbrich, S., Optimization with PDE Constraints, Springer, New York, (2009)
20. Ji, X., Sun, J., Yang, Y., Optimal penalty parameter for C^0 IPDG. Appl. Math. Lett., **37**, 112–117, (2014)
21. Kärkkäinen, T., Kunisch, K., Tarvainen, P., Augmented Lagrangian active set methods for obstacle problems. J. Optim. Theory Appl., **119**, 499–533 (2003)
22. Kinderlehrer, D., Stampacchia, G., An Introduction to Variational Inequalities and Their Applications. Society for Industrial and Applied Mathematics, Philadelphia, (2000)
23. Lions, J.-L., Stampacchia, G., Variational inequalities. Comm. Pure Appl. Math., **20**, 493–519, (1967)
24. Liu, W., Gong, W., Yan, N., A new finite element approximation of a state-constrained optimal control problem. J. Comput. Math., **27**, 97–114, (2009)
25. Maz'ya, V., Rossmann, J., Elliptic Equations in Polyhedral Domains. American Mathematical Society, Providence, RI, (2010)
26. Rodrigues, J.-F., Obstacle Problems in Mathematical Physics. North-Holland Publishing Co., Amsterdam, **134**, (1987)

The Effect of the Sensitivity Parameter in Weighted Essentially Non-oscillatory Methods

Bo Dong, Sigal Gottlieb, Yulia Hristova, Yan Jiang and Haijin Wang

Abstract Weighted essentially non-oscillatory methods (WENO) were developed to capture shocks in the solution of hyperbolic conservation laws while maintaining stability and without smearing the shock profile. WENO methods accomplish this by assigning weights to a number of candidate stencils, according to the smoothness of the solution on the stencil. These weights favor smoother stencils when there is a significant difference while combining all the stencils to attain higher order when the stencils are all smooth. When WENO methods were initially introduced, a small parameter ε was defined to avoid division by zero. Over time, it has become apparent that ε plays the role of the sensitivity parameter in stencil selection. WENO methods allow some oscillations, and it is well known that these oscillations depend on the size of ε. In this work, we show that the value of ε must be below a certain critical threshold ε_c and that this threshold depends on the function used and on the size of the jump discontinuity captured. Next, we analytically and numerically show the size of the oscillations for one time-step and over long time integration when $\varepsilon < \varepsilon_c$

B. Dong · S. Gottlieb (✉)
Department of Mathematics, University of Massachusetts Dartmouth,
Dartmouth, MA, USA
e-mail: sgottlieb@umassd.edu

B. Dong
e-mail: bdong@umassd.edu

Y. Hristova
Department of Mathematics and Statistics, University of Michigan-Dearborn,
Dearborn, MI, USA
e-mail: yuliagh@umich.edu

Y. Jiang
Department of Mathematics, Michigan State University, 619 Red Cedar Rd.,
East Lansing 48824, MI, USA
e-mail: jiangyan@math.msu.edu

H. Wang
College of Science, Nanjing University of Posts and Telecommunications, Nanjing 210023,
Jiangsu Province, People's Republic of China
e-mail: hjwang@njupt.edu.cn

© Springer Science+Business Media New York 2016
S.C. Brenner (ed.), *Topics in Numerical Partial Differential Equations and Scientific Computing*, The IMA Volumes in Mathematics and its Applications 160, DOI 10.1007/978-1-4939-6399-7_2

and their dependence on the size of ε, the function used, and the size of the jump discontinuity.

1 Introduction: Weighted Essentially Non-oscillatory Methods

When approximating the solution to a conservation law of the form

$$u_t + f(u)_x = 0,$$

we use a conservative finite difference scheme

$$u_t = -\frac{1}{\Delta x}(\hat{f}_{j+\frac{1}{2}} - \hat{f}_{j-\frac{1}{2}})$$

to obtain a physically relevant solution [9]. The term $\hat{f}_{j+\frac{1}{2}} = \hat{f}(u_{j-k}, \ldots, u_{j+l})$ is the *numerical flux*, and the points $x_{j-k}, \ldots, x_{j+l}$ constitute the *stencil*. To be a reasonable approximation, the numerical flux must be (at least) Lipschitz continuous and consistent with the physical flux f, i.e., $\hat{f}(u, \ldots, u) = f(u)$. Once the spatial derivative is computed by differencing the numerical fluxes, we obtain a system of ODEs which is then evolved to the next time-step using some standard time-stepping method. Different numerical fluxes give rise to different numerical methods. Any differences between the properties of such methods are a result of differences in the numerical flux.

A major issue with the use of finite difference methods for computations with shocks is that oscillations arise when we take points on opposite sides of the shock to evaluate the derivative at a given point. These oscillations at the shock location propagate to the smooth regions, destroying the stability of the solution. To avoid oscillations and instabilities that arise from using a finite-difference stencil that takes information from both sides of the shock, ENO ([3], [4], [12]) schemes search for the locally smoothest stencil and use that stencil to calculate the numerical fluxes. The idea behind ENO schemes is stencil switching in order to eliminate oscillations. The ENO scheme evaluates the smoothness of several stencils near the point of interest and picks the smoothest stencil for evaluating the derivative at that point. This means that the method should avoid picking stencils that cross the shock, so that the stencil is chosen only from a smooth region (to the left or right of the shock) in which linear stability is enough to ensure nonlinear stability.

Liu, Osher, and Chan [10] improved upon the ENO method by assigning each stencil a weight which depends on its smoothness and summing the approximations from all the candidate stencils, each with its corresponding weight. The weights are chosen so that in smooth regions, we obtain higher-order accuracy, whereas near discontinuities, the method imitates the ENO scheme by assigning near-zero weights to the stencils that contain discontinuities. This approach is called the weighted ENO (or WENO) method. An rth order ENO scheme considers a total of $2r - 1$ points to evaluate the flux. The WENO scheme uses all the candidate stencils and therefore $2r - 1$ points, so that a clever choice of weights [7] results in a WENO scheme which is of order $2r - 1$ in smooth regions [13].

In the following subsections, we describe three variants of the WENO procedure. We focus on the choice $r = 3$ which gives fifth-order methods for smooth problems and third order in non-smooth regions. The three methods we consider are the WENO method described by Jiang and Shu in [7], the mapped WENO procedure given in [5], and the improved method presented in [1]. We refer to these methods as WENO-JS, WENO-M, and WENO-Z, respectively. In all of the following, we assume that we have a flux $f(u)$ such that $\frac{df}{du} \geq 0$. If this is not the case, then we split the flux into the positive and negative parts

$$f(u) = f^+(u) + f^-(u),$$

such that $\frac{df^+}{du} \geq 0$ and $\frac{df^-}{du} \leq 0$. Then, we handle each part separately, using differently biased stencils for the negative flux. We will not describe this in detail here, but the interested reader can consult [13].

1.1 WENO-JS

In this section, we present the $r = 3$ method of Jiang and Shu [7]. To calculate the numerical flux $\hat{f}_{j+\frac{1}{2}}$ at any given point x_j, we begin by calculating the smoothness measurements to determine whether a shock lies within the stencil. For the fifth-order scheme, these are as follows:

$$IS_0 = \frac{13}{12}\left(f_{j-2} - 2f_{j-1} + f_j\right)^2 + \frac{1}{4}\left(f_{j-2} - 4f_{j-1} + 3f_j\right)^2$$
$$IS_1 = \frac{13}{12}\left(f_{j-1} - 2f_j + f_{j+1}\right)^2 + \frac{1}{4}\left(f_{j-1} - f_{j+1}\right)^2$$
$$IS_2 = \frac{13}{12}\left(f_j - 2f_{j+1} + f_{j+2}\right)^2 + \frac{1}{4}\left(3f_j - 4f_{j+1} + f_{j+2}\right)^2$$

(Note that the factor of $\frac{1}{12}$ can be ignored due to the normalization later). Next, we use the smoothness measurements to calculate the stencil weights

$$\alpha_0^{(JS)} = \omega_0\left(\frac{1}{\varepsilon + IS_0}\right)^p \qquad \alpha_1^{(JS)} = \omega_1\left(\frac{1}{\varepsilon + IS_1}\right)^p \qquad \alpha_2^{(JS)} = \omega_2\left(\frac{1}{\varepsilon + IS_2}\right)^p$$

where $\omega_0 = \frac{1}{10}$, $\omega_1 = \frac{6}{10}$, and $\omega_2 = \frac{3}{10}$ are the centered difference weights, and p is the power of the weights, typically chosen to be $p = 2$.

Note that the parameter ε is added to the denominator to prevent division by zero. To avoid division by machine zero, we must pick ε larger than the square root of the smallest positive number seen as nonzero by the computer. For single precision, this is $\varepsilon > 10^{-18}$, while for double precision, we need $\varepsilon > 10^{-153}$.

These weights are then normalized

$$\omega_0^{(JS)} = \frac{\alpha_0^{(JS)}}{\sum_{i=1}^3 \alpha_i^{(JS)}} \qquad \omega_1^{(JS)} = \frac{\alpha_1^{(JS)}}{\sum_{i=1}^3 \alpha_i^{(JS)}} \qquad \omega_2^{(JS)} = \frac{\alpha_2^{(JS)}}{\sum_{i=1}^3 \alpha_i^{(JS)}},$$

and the normalized weights are used to compute the numerical fluxes

$$\hat{f}_{j+\frac{1}{2}} = \omega_0^{(JS)} \left(\frac{2}{6} f_{j-2} - \frac{7}{6} f_{j-1} + \frac{11}{6} f_j \right) + \omega_1^{(JS)} \left(-\frac{1}{6} f_{j-1} + \frac{5}{6} f_j + \frac{2}{6} f_{j+1} \right)$$
$$+ \omega_2^{(JS)} \left(\frac{2}{6} f_j + \frac{5}{6} f_{j+1} - \frac{1}{6} f_{j+2} \right).$$

Finally, the derivative is computed by taking a difference of the fluxes

$$f(u)_x \approx \frac{1}{\Delta x} \left(\hat{f}_{j+\frac{1}{2}}^+ - \hat{f}_{j-\frac{1}{2}}^+ \right).$$

This process leads to a system of ordinary differential equations, which can then be evolved by standard time-stepping methods such as Runge–Kutta schemes.

1.2 Mapped WENO (WENO-M)

The mapped WENO method [5] was developed to overcome the loss of order of convergence near critical points of f that is experienced by WENO-JS. Rather than creating a new smoothness measure, the mapped WENO algorithm uses the ideal weights ω_k and the WENO-JS weights $\omega_k^{(JS)}$ to create new mapped weights, $\omega_k^{(M)}$ given by

$$\omega_k^{(M)} = \frac{\alpha_k^{(M)}}{\sum_{i=0}^2 \alpha_i^{(M)}}$$

where

$$\alpha_k^{(M)} = \frac{\omega_k^{(JS)} \left(\omega_k + \omega_k^2 - 3\omega_k \omega_k^{(JS)} + (\omega_k^{(JS)})^2 \right)}{\omega_k^2 + \omega_k^{(JS)}(1 - 2\omega_k)}.$$

These weights are then used for the computation of the numerical fluxes

$$\hat{f}_{j+\frac{1}{2}} = \omega_0^{(M)} \left(\frac{2}{6} f_{j-2} - \frac{7}{6} f_{j-1} + \frac{11}{6} f_j \right) + \omega_1^{(M)} \left(-\frac{1}{6} f_{j-1} + \frac{5}{6} f_j + \frac{2}{6} f_{j+1} \right)$$
$$+ \omega_2^{(M)} \left(\frac{2}{6} f_j + \frac{5}{6} f_{j+1} - \frac{1}{6} f_{j+2} \right).$$

The convergence of this WENO-M method near critical points was studied in [5] with the value $\varepsilon = 10^{-40}$ and verified that this method converges at design order even near critical points.

1.3 WENO-Z

The mapped WENO approach above is much slower than the original WENO-JS. An improvement of both these methods was introduced in [1] where the smoothness measurements are modified. This method defines a value $\tau_5 = |IS_0 - IS_2|$ and computes the new weights

$$\alpha_k^{(z)} = \omega_k \left(1 + \left(\frac{\tau_5}{IS_k + \varepsilon} \right)^p \right).$$

These weights are then normalized

$$\omega_0^{(z)} = \frac{\alpha_0^{(z)}}{\sum_{i=1}^3 \alpha_i^{(z)}} \qquad \omega_1^{(z)} = \frac{\alpha_1^{(z)}}{\sum_{i=1}^3 \alpha_i^{(z)}} \qquad \omega_2^{(z)} = \frac{\alpha_2^{(z)}}{\sum_{i=1}^3 \alpha_i^{(z)}},$$

and the normalized weights are used to compute the numerical fluxes

$$\hat{f}_{j+\frac{1}{2}} = \omega_0^{(z)} \left(\frac{2}{6} f_{j-2} - \frac{7}{6} f_{j-1} + \frac{11}{6} f_j \right) + \omega_1^{(z)} \left(-\frac{1}{6} f_{j-1} + \frac{5}{6} f_j + \frac{2}{6} f_{j+1} \right)$$
$$+ \omega_2^{(z)} \left(\frac{2}{6} f_j + \frac{5}{6} f_{j+1} - \frac{1}{6} f_{j+2} \right).$$

The fifth-order WENO-Z scheme was shown to have less dissipation and higher resolution than the WENO-JS scheme, generates solutions that are as sharp as those in WENO-M, and does not suffer as much as WENO-JS from reduced convergence rates at critical points. In addition, the WENO-Z method has a computational cost about the same as WENO-JS and significantly smaller than WENO-M.

In [1], the authors consistently use a small value of ε to avoid this parameter dominating over the smoothness measurements. Later, in [6], the authors consider the appropriate values of ε and suggest the value $\varepsilon = \Delta x^{r-1}$ as a compromise that avoids the loss of stencil sensitivity when ε dominates over the smoothness measurement, while still serving its original role of preventing division by zero.

2 The Importance of Choosing a Small Enough Sensitivity Parameter

When the parameter ε was initially introduced, its sole function was to prevent division by zero. The parameter value chosen initially was $\varepsilon = 10^{-6}$ [7]. In fact, the parameter ε serves a role as a sensitivity parameter that determines the difference in size between smoothness measurements that induce stencil switching. If the smoothness indicators are all below the level of ε, all the candidate stencils are seen as equally smooth and the centered difference method is attained. However, if one of the smoothness indicators is larger than ε, then the weights will no longer be equal and the chosen stencils will be biased away from the shock.

Over time, users of the WENO method have observed that the value of ε had important implications on the presence of oscillations. In studying the order of convergence of WENO-JS, Henrick et al. [5] noted that the parameter ε plays a critical though unintended role. They noted that the expansion of the nonnormalized weights yields $\alpha_k = \frac{\omega_k}{\varepsilon + \Delta x^2 f'^2 + O(\Delta x^4)}$ so that as $\Delta x \to 0$, the parameter ε will eventually dominate over the smoothness indicators, and the WENO-JS method approaches the behavior of the central difference scheme. At the same time, they also note that oscillations of order ε^2 can be seen near discontinuities (an observation that we will study rigorously later in this paper). Both these observations suggest that smaller values of ε are preferable. The first question we ask is what is that value of ε_c, the critical value that ε must be below so that its effect is easily controlled. In this section, we perform a few numerical studies that show the dependence of ε_c on the power of the function and the size of the jump.

To understand what is happening, we start with a simple example: consider the function $f(u) = cu^m$ where

$$u = \begin{cases} 0 & \text{if } n \leq j \\ M & \text{if } n > j \end{cases}$$

In this case, m is the order of the function, and M is the size of the jump. The smoothness measurements for this function are mostly zero, except near the shock. At that point, we get

$$IS_0 = \frac{13}{12} \left(f_{j-2} - 2f_{j-1} + f_j \right)^2 + \frac{1}{4} \left(f_{j-2} - 4f_{j-1} + 3f_j \right)^2 = 0$$

$$IS_1 = \frac{13}{12} \left(f_{j-1} - 2f_j + f_{j+1} \right)^2 + \frac{1}{4} \left(f_{j-1} - f_{j+1} \right)^2 = \frac{4}{3} c^2 (M)^{2m}$$

$$IS_2 = \frac{13}{12} \left(f_j - 2f_{j+1} + f_{j+2} \right)^2 + \frac{1}{4} \left(3f_j - 4f_{j+1} + f_{j+2} \right)^2 = \frac{10}{3} c^2 (M)^{2m}.$$

The weights then become (with the choice of $p = 2$)

$$\alpha_0 = \frac{1}{10} \left(\frac{1}{\varepsilon} \right)^2, \quad \alpha_1 = \frac{6}{10} \left(\frac{1}{\varepsilon + \frac{4}{3} c^2 (M)^{2m}} \right)^2, \quad \alpha_2 = \frac{3}{10} \left(\frac{1}{\varepsilon + \frac{10}{3} c^2 (M)^{2m}} \right)^2.$$

Notice that if $\varepsilon \ll c^2 M^{2m}$, the smooth stencil will get the large weights and the other stencils get very small weights. However, if $\varepsilon \geq \varepsilon_c \approx c^2 M^{2m}$, the wrong stencils will be chosen, producing oscillations. The purpose of this section is to explore this critical value of ε and its effects in a variety of numerical settings.

In the following examples, we demonstrate the effect of choosing a value of ε that is too large and examine the values of ε that are small enough to prevent unwanted behavior of the method. First, we examine the size of the overshoot/undershoot in a simple linear advection equation.

Example 1: Consider linear advection equation

$$u_t + u_x = 0 \quad \text{with} \quad u(x, 0) = \begin{cases} -1 & \text{if } x < 0 \\ 1 & \text{otherwise} \end{cases}$$

We discretize the spatial grid with $N = 100$ points between $[-1, 1]$. We use WENO-JS to evaluate the derivatives and use the third-order SSP Runge–Kutta method with $\Delta t = \frac{1}{2}\Delta x$ to advect the solution over fifty time-steps. If $|u| > 1$, we have an undershoot or overshoot. The table below lists the maximal over/undershoot for the first 50 time-steps for each value of ε. We observe that for this example, we require $\varepsilon \leq 10^{-24}$ for the under- or overshoot to be within roundoff error and even smaller ε for the under- or overshoot to be within machine precision.

ε	10^{-6}	10^{-9}	10^{-13}	10^{-16}	10^{-19}	10^{-24}	10^{-29}	10^{-39}
$max|u| - 1$	5.66e-5	1.62e-6	1.42e-8	4.16e-10	1.18e-11	3.26e-14	2.22e-16	2.22e-16

This simple demonstration indicates that the choice of ε strongly affects the size of the oscillations. There is a critical value ε_c below which the oscillations are no longer significant. In the next section, we explore the dependence of this critical value of ε on the size of the jump.

2.1 Dependence of the Critical Value of ε on the Size of the Jump Discontinuity

In this section, we study numerically the behavior of several problems, with several jump sizes, using the WENO-JS method.

Example 1a We revisit the linear advection equation in **Example 1**, this time with step function initial conditions of size M

$$u(x, 0) = \begin{cases} M & \text{if } -1/8 < x < 1/8 \\ 0 & \text{otherwise} \end{cases}$$

If we use a relatively large $\varepsilon = M = 10^{-2}$, we get very bad looking results after 50 time-steps, as seen in Figure 1.

Fig. 1 Linear advection with $\varepsilon = M = 10^{-2}$, WENO-JS.

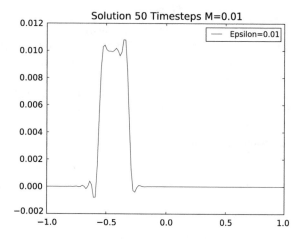

But even if the results look nice, ε can have a bad effect on the total variation (TV) of the numerical solution as seen in Figure 2 on the left, with $M = 1$. Another observation is that the smaller M is, the smaller we require ε to avoid a significant TV increase (Figure 2 with $M = 10^{-1}$ and $M = 10^{-4}$). Although the TV of this problem increases generally, whatever the value of ε, an additional rise that is attributable to the value of ε is concerning. In these simulations, we observe that the critical values of ε needed to prevent a spurious rise in total variation are $\varepsilon_c \approx 10^{-8}M^2$.

Example 2 We consider Burgers' equation

$$u_t + \left(\frac{1}{2}u^2\right)_x = 0 \qquad x \in (-1, 1)$$

with initial condition

$$u(x, 0) = M \sin(\pi x).$$

This problem develops a stationary shock. We discretize this grid with 100 points in space and set $\Delta t = \frac{CFL}{M}\Delta x$ with $CFL = 0.5$. Choosing $M = 0.01$ and $\varepsilon = 10^{-5}$, we step this forward for 50 time-steps. The results show an oscillatory profile, as seen in Figure 3. If ε is chosen small enough, there is no rise in TV. The values of ε_c for each M, required for TV norm to settle, are seen in the table below.

M	10^{-1}	10^{-3}	10^{-5}	10^{-6}	$10^{-6.5}$	10^{-7}	10^{-9}
ε_c	10^{-5}	10^{-13}	10^{-21}	10^{-25}	10^{-27}	10^{-29}	10^{-37}

$$\longrightarrow \quad \varepsilon_c \approx 10^{-1}M^4.$$

Example 3 To see whether the pattern of dependence on m continues, we test the problem

$$u_t + \left(\frac{1}{3}u^3\right)_x = 0 \quad \text{with} \quad u(x, 0) = \begin{cases} M & \text{if } -1/8 < x < 1/8 \\ 0 & \text{otherwise} \end{cases}$$

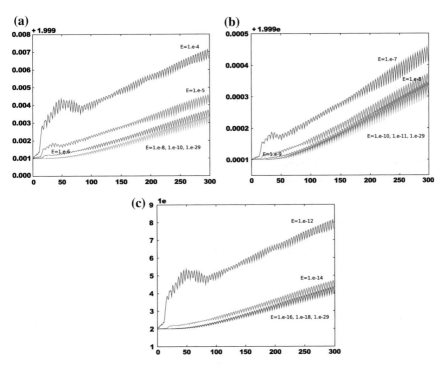

Fig. 2 Linear advection with different values of M. The smaller the M is, the smaller ε needs to be to avoid a significant TV increase. (a) $M = 1$. (b) $M = 10^{-1}$. (c) $M = 10^{-4}$.

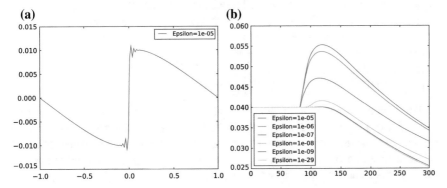

Fig. 3 Burgers' example with $M = 10^{-2}$. Left: numerical solution for $\varepsilon = 10^{-5}$, after 50 time-steps. Right: TV of the solution over 300 time-steps, for different values of ε. (a) Solution Profile with $\varepsilon = 10^{-5}$. (b) Total Variation.

The relationship between the value of M and the ε required to avoid a TV "bump" is shown in the table below.

M	10^{-0}	10^{-1}	10^{-2}	10^{-3}	10^{-4}	10^{-5}	10^{-6}
ε_c	10^{-3}	10^{-9}	10^{-15}	10^{-21}	10^{-27}	10^{-33}	10^{-39}

$$\longrightarrow \quad \varepsilon_c \approx 10^{-3}M^6.$$

Example 4 Consider the Euler system

$$\begin{pmatrix} \rho \\ \rho u \\ E \end{pmatrix} + \begin{pmatrix} \rho u \\ P + \rho u^2 \\ u(P + E) \end{pmatrix} = 0$$

on $0 \le x \le 9$, and $t \ge 0$. Here, $\rho(x, t)$ is the density, $\rho u(x, t)$ the momentum, $E(x, t)$ the energy, $P(x, t)$ is the pressure, and $c(x, t)$ is the soundspeed. They are related by $P(x, t) = (\gamma - 1)\left(E - \frac{1}{2}\rho v^2\right)$, and $c(x, t) = \sqrt{\frac{\gamma P}{\rho}}$.

We impose boundary conditions

$$\rho_L = M \quad u_L = 0 \quad P_L = M * (\gamma - 1) * 0.1$$

$$\rho_R = 0.001M \quad u_R = 0 \quad P_R = M * (\gamma - 1) * 10^{-7}$$

and step function initial conditions

$$\rho(x, t) = \begin{cases} \rho_L & \text{if } x \le 3 \\ \rho_R & \text{if } x > 3 \end{cases} \quad u(x, t) = \begin{cases} u_L & \text{if } x \le 3 \\ u_R & \text{if } x > 3 \end{cases} \quad P(x, t) = \begin{cases} P_L & \text{if } x \le 3 \\ P_R & \text{if } x > 3. \end{cases}$$

For the Euler system, it is important that the pressure or density does not become negative, even at intermediate stages. The value of ε has an effect on this, as well. The table below shows the values of ε_c such that when $\varepsilon > \varepsilon_c$, we get negative pressure or density and the code stops working.

M	1	10^{-2}	10^{-4}	10^{-6}	10^{-8}	10^{-10}	10^{-12}
ε_c	10^{-5}	10^{-9}	10^{-13}	10^{-17}	10^{-21}	10^{-25}	10^{-29}

$$\longrightarrow \quad \varepsilon_c \approx 10^{-5}M^2.$$

This section provided numerical evidence that we must choose $\varepsilon \le \varepsilon_c \approx M^{2m}$ to avoid a variety of problems is $\varepsilon_c \approx M^{2m}$. The important observation here is that the size of ε varies with the size of the jump. In the next section, we consider the behavior of the method when $\varepsilon \ll \varepsilon_c$ and observe the size of the oscillations that occur in that range and their dependence on ε.

3   The Effect of the Sensitivity Parameter for One Time-step

In the previous section, we showed that ε functions as a stencil sensitivity parameter and that the value of ε needs to be chosen appropriately to avoid or mitigate the effect of undershoots or overshoots. The presence of an undershoot or overshoot even in a single time-step can have a profound effect on the computation. In the

Euler system example above, we saw that case where pressure or density becomes negative, and the code will no longer work. Setting these quantities to zero when they become negative resolves this problem, but may introduce significant errors into the computation. Thus, the aim is to avoid undershoots and overshoots as much as possible. In this section, we will see that any nonzero ε values may produce oscillations and discuss the dependence of these oscillations on ε as well as the size of the jump M. We will also examine the effect of the ratio of the time-step to the spatial grid size plays a role in the size of the oscillations.

In the following, we consider the prototype problem

$$u_t + f(u)_x = 0 \tag{1}$$

with step-function initial conditions

$$u(x, 0) = \begin{cases} M, & x \in [-\frac{1}{8}, \frac{1}{8}] \\ 0, & otherwise \end{cases} \tag{2}$$

where $f(u) = \frac{1}{m}u^m$ for $m = 1, 2, 3, 4$. This problem is a representative of all Riemann problems, which are the building blocks of methods for hyperbolic PDEs.

In the previous section, we looked at what happens if ε is too large for a particular stencil. In this section, we study small ε sizes and determine the effect of the size of ε on the presence of oscillations. We begin with a mathematical analysis of the behavior of the undershoot for the example above with WENO-JS for one time-step using forward Euler (FE) and the two-stage second-order SSP Runge–Kutta method (RK2) and verify this analysis with numerical simulations. We then proceed to show numerically the behavior of the oscillations for one time-step of the three-stage third-order SSP Runge–Kutta method (RK3,3) and for the ten-stage fourth-order SSP Runge–Kutta method (RK10,4). We repeat these numerical explorations for WENO-M and for WENO-Z and observe similar behaviors. Our conclusion in this section is that size of the oscillation after one time-step scales as

$$\text{oscillations} \approx \frac{\varepsilon^2}{M^{4m-1}},$$

where this scaling depends on the CFL, the particular WENO method, and the value of m. It is important to note that smaller jumps M result in larger oscillations, all else being equal.

3.1 One Step of WENO-JS with Forward Euler Timestepping

FE is the first stage of many higher-order Runge–Kutta methods. A thorough study of WENO-JS with FE will help us have a better understanding of the behavior of WENO methods with other Runge–Kutta methods. In this section, we first estimate the undershoot/overshoot error of WENO-JS with FE for the linear case $f(u) = u$

and then extend the analysis to the nonlinear case $f(u) = \frac{1}{m}u^m$ where $m = 2, 3, 4$. We also show numerical results, which are consistent with our theoretical estimates.

3.1.1 Linear Case

First, we consider the behavior of WENO-JS on problem (1) above with $m = 1$ and initial conditions (2). We assume that $\varepsilon << M$ and examine the values of the numerical flux $\hat{f}_{j+1/2}$ for each possible j in a case-by-case sense. Note that in the formulas for the smoothness indicators, we ignore the factor of $\frac{1}{12}$ multiplying every term:

Stencil values   numerical flux

$u_{j-2}, \ldots, u_{j+2} = 0 \quad \Rightarrow \quad \hat{f}_{j+1/2} = 0$

$u_{j-2}, \ldots, u_{j+1} = 0$ and $u_{j+2} = M \quad \Rightarrow \quad \hat{f}_{j+1/2} = \frac{\frac{3}{(16M^2+\varepsilon)^2}}{\frac{7}{\varepsilon^2} + \frac{3}{(16M^2+\varepsilon)^2}}(-\frac{1}{6}M)$

$\qquad\qquad\qquad\qquad\qquad\qquad \approx -\frac{1}{3584}(\frac{\varepsilon}{M^2})^2 M$

$u_{j-2}, \ldots, u_j = 0$ and $u_{j+1}, u_{j+2} = M \quad \Rightarrow \quad \hat{f}_{j+1/2} = \frac{\frac{6}{(16M^2+\varepsilon)^2}(\frac{2}{6}M) + \frac{3}{(40M^2+\varepsilon)^2}(\frac{4}{6}M)}{\frac{1}{\varepsilon^2} + \frac{6}{(16M^2+\varepsilon)^2} + \frac{3}{(40M^2+\varepsilon)^2}}$

$\qquad\qquad\qquad\qquad\qquad\qquad \approx \frac{29}{3200}(\frac{\varepsilon}{M^2})^2 M$

$u_{j-2}, u_{j-1} = 0$ and $u_j, \ldots, u_{j+2} = M \quad \Rightarrow \quad \hat{f}_{j+1/2} = \frac{\frac{1}{(40M^2+\varepsilon)^2}(\frac{11}{6}M) + \frac{6}{(16M^2+\varepsilon)^2}(\frac{7}{6}M) + \frac{3}{\varepsilon^2}M}{\frac{1}{(40M^2+\varepsilon)^2} + \frac{6}{(16M^2+\varepsilon)^2} + \frac{3}{\varepsilon^2}}$

$\qquad\qquad\qquad\qquad\qquad\qquad \approx (1 + \frac{17}{11520}(\frac{\varepsilon}{M^2})^2)M$

$u_{j-2} = 0$ and $u_{j-1}, \ldots, u_{j+2} = M \quad \Rightarrow \quad \hat{f}_{j+1/2} = \frac{\frac{1}{(16M^2+\varepsilon)^2}(\frac{4}{6}M) + \frac{6}{\varepsilon^2}M + \frac{3}{\varepsilon^2}M}{\frac{1}{(16M^2+\varepsilon)^2} + \frac{6}{\varepsilon^2} + \frac{3}{\varepsilon^2}}$

$\qquad\qquad\qquad\qquad\qquad\qquad \approx (1 - \frac{1}{6912}(\frac{\varepsilon}{M^2})^2)M$

$u_{j-2}, \ldots, u_{j+2} = M \quad \Rightarrow \quad \hat{f}_{j+1/2} = M$

Now, we consider what happens near the first discontinuity, the jump from 0 to M, when using the WENO-JS method with one step of the first-order forward Euler (FE) method in time:

$$u_j^1 = u_j^0 - \lambda(\hat{f}_{j+1/2} - \hat{f}_{j-1/2}),$$

where $\lambda = \frac{\Delta t}{\Delta x}$.

1. If $u_{j-3}^0, \ldots, u_{j+2}^0 = 0$, then $u_j^1 = 0$
2. If $u_{j-3}^0, \ldots, u_{j+1}^0 = 0$ and $u_{j+2}^0 = M$, then $u_j^1 \approx -\lambda\{-\frac{1}{3584}(\frac{\varepsilon}{M^2})^2 M\}$
   $= \frac{1}{3584}\lambda(\frac{\varepsilon}{M^2})^2 M$
3. If $u_{j-3}^0, \ldots, u_j^0 = 0$ and $u_{j+1}^0, u_{j+2}^0 = M$, then $u_j^1 \approx -\lambda\{\frac{29}{3200}(\frac{\varepsilon}{M^2})^2 M$
   $+ \frac{1}{3584}(\frac{\varepsilon}{M^2})^2 M\} = -\frac{837}{89600}\lambda(\frac{\varepsilon}{M^2})^2 M$ and there will be undershoot.
4. If $u_{j-3}^0, \ldots, u_{j-1}^0 = 0$ and $u_j^0, \ldots, u_{j+2}^0 = M$, then
   $u_j^1 \approx M - \lambda\{(1 + \frac{17}{11520}(\frac{\varepsilon}{M^2})^2)M - \frac{29}{3200}(\frac{\varepsilon}{M^2})^2 M\} = \{1 - \lambda(1 - \frac{437}{57600}(\frac{\varepsilon}{M^2})^2)\}M$
5. If $u_{j-3}^0, u_{j-2}^0 = 0$ and $u_{j-1}^0, \ldots, u_{j+2}^0 = M$, then
   $u_j^1 \approx M - \lambda\{(1 - \frac{1}{6912}(\frac{\varepsilon}{M^2})^2)M - (1 + \frac{17}{11520}(\frac{\varepsilon}{M^2})^2)M\} = \{1 + \frac{7}{4320}\lambda(\frac{\varepsilon}{M^2})^2\}M$
   and there will be overshoot.

6. If $u_{j-3}^0 = 0$ and $u_{j-2}^0, \ldots, u_{j+2}^0 = M$, then $u_j^1 \approx M - \lambda\{M - (1 - \frac{1}{6912}(\frac{\varepsilon}{M^2})^2)M\} = \{1 - \frac{1}{6912}\lambda(\frac{\varepsilon}{M^2})^2\}M$
7. If $u_{j-3}^0, \ldots, u_{j+2}^0 = M$, then $u_j^1 = M$.

In the same way, we can also compute u_j^1 when u_j^0 jump from M to 0:

1. If $u_{j-3}^0, \ldots, u_{j+1}^0 = M$ and $u_{j+2}^0 = 0$, then $u_j^1 \approx M - \lambda\{(1 + \frac{1}{3584}(\frac{\varepsilon}{M^2})^2)M - M\} = \{1 - \frac{1}{3584}\lambda(\frac{\varepsilon}{M^2})^2\}M$.
2. If $u_{j-3}^0, \ldots, u_j^0 = M$ and $u_{j+1}^0, u_{j+2}^0 = 0$, then
   $$u_j^1 \approx M - \lambda\{(1 - \frac{29}{3200}(\frac{\varepsilon}{M^2})^2)M - (1 + \frac{1}{3584}(\frac{\varepsilon}{M^2})^2)M\} = \{1 + \frac{837}{89600}\lambda(\frac{\varepsilon}{M^2})^2\}M$$
   and there will be overshoot.
3. If $u_{j-3}^0, \ldots, u_{j-1}^0 = M$ and $u_j^0, \ldots, u_{j+2}^0 = 0$, then
   $$u_j^1 \approx -\lambda\{-\frac{17}{11520}(\frac{\varepsilon}{M^2})^2 M - (1 - \frac{29}{3200}(\frac{\varepsilon}{M^2})^2)M\} = \lambda\{1 - \frac{437}{57600}(\frac{\varepsilon}{M^2})^2\}M$$
4. If $u_{j-3}^0, u_{j-2}^0 = M$ and $u_{j-1}^0, \ldots, u_{j+2}^0 = 0$, then $u_j^1 \approx -\lambda\{\frac{1}{6912}(\frac{\varepsilon}{M^2})^2 M + \frac{17}{11520}(\frac{\varepsilon}{M^2})^2 M\} = -\frac{7}{4320}\lambda(\frac{\varepsilon}{M^2})^2 M$ and there will be undershoot.
5. If $u_{j-3}^0 = M$ and $u_{j-2}^0, \ldots, u_{j+2}^0 = 0$, then $u_j^1 \approx -\lambda\{0 - \frac{1}{6912}(\frac{\varepsilon}{M^2})^2 M\} = \frac{1}{6912}\lambda(\frac{\varepsilon}{M^2})^2 M$.

In summary, after one forward Euler time-step of WENO-JS, the maximum of overshoot is $\lambda\frac{837}{89600}(\frac{\varepsilon}{M^2})^2 M = \lambda\frac{837}{89600}\frac{\varepsilon^2}{M^3}$, and the maximum of undershoot is $-\lambda\frac{837}{89600}\frac{\varepsilon^2}{M^3}$. In Table 1, we show numerically that the actual undershoot error of WENO-JS after one forward Euler time-step with $\lambda = 0.25$ is $2.3354 \times 10^{-3}\frac{\varepsilon^2}{M^3}$,

Table 1 The undershoot error of WENO-JS after one time-step with FE, using step function initial conditions, with $nx = 200$ points in space, and $CFL = 0.25$.

		$M = 1$	$M = 0.1$	$M = 0.01$	$M = 0.001$	pattern
$f(u) = u$	$\varepsilon = 1e\text{-}10$	2.3354e-23	2.3354e-20	2.3354e-17	2.3354e-14	$2.3354 \times 10^{-3}\frac{\varepsilon^2}{M^3}$
	$\varepsilon = 1e\text{-}11$	2.3354e-25	2.3354e-22	2.3354e-19	2.3354e-16	
	$\varepsilon = 1e\text{-}12$	2.3354e-27	2.3354e-24	2.3354e-21	2.3354e-18	
$f(u) = u^2/2$	$\varepsilon = 1e\text{-}18$	2.5761e-38	2.5761e-31	2.5761e-24	2.5761e-17	$2.5761 \times 10^{-2}\frac{\varepsilon^2}{M^7}$
	$\varepsilon = 1e\text{-}19$	2.5761e-40	2.5761e-33	2.5761e-26	2.5761e-19	
	$\varepsilon = 1e\text{-}20$	2.5761e-42	2.5761e-35	2.5761e-28	2.5761e-21	
$f(u) = u^3/3$	$\varepsilon = 1e\text{-}23$	1.0702e-48	1.0702e-37	1.0702e-26	1.0702e-15	$1.0702 \times 10^{-2}\frac{\varepsilon^2}{M^{11}}$
	$\varepsilon = 1e\text{-}24$	1.0702e-50	1.0702e-39	1.0702e-28	1.0702e-17	
	$\varepsilon = 1e\text{-}25$	1.0702e-52	1.0702e-41	1.0702e-30	1.0702e-19	
$f(u) = u^4/4$	$\varepsilon = 1e\text{-}29$	7.3960e-61	7.3960e-46	7.3960e-31	7.3960e-16	$7.3960 \times 10^{-3}\frac{\varepsilon^2}{M^{15}}$
	$\varepsilon = 1e\text{-}30$	7.3960e-63	7.3960e-48	7.3960e-33	7.3960e-18	
	$\varepsilon = 1e\text{-}31$	7.3960e-65	7.3960e-50	7.3960e-35	7.3960e-20	

Table 2 The undershoot error of WENO-JS after one time-step with FE using step function initial conditions, $nx = 200, M = 1, \varepsilon = 1e - 10$. This table shows the linear increase with the CFL number.

	$CFL = 0.125$	$CFL = 0.25$	$CFL = 0.5$
$f(u) = u$	1.1677e-23	2.3354e-23	4.6708e-23
$f(u) = u^2/2$	1.2880e-22	2.5761e-22	5.1521e-22
$f(u) = u^3/3$	5.3510e-23	1.0702e-22	2.1404e-22
$f(u) = u^4/4$	3.6980e-23	7.3960e-23	1.4792e-22

which is *exactly* the same as the undershoot/overshoot value

$$\lambda \frac{837}{89600} \frac{\varepsilon^2}{M^3} \tag{3}$$

derived above. This shows that the analysis above and the resulting estimate of the undershoot error for the case where ε is small enough is *sharp*. The numerical results in Table 2 confirm that the undershoot error is proportional to the CFL number, which is equal to λ in the linear case.

3.1.2 Nonlinear Cases

The analysis of the undershoot and overshoot in the case where f is nonlinear ($m > 1$) can be estimated in the same way as that for the linear case in the section above. However, for the nonlinear case, calculations are long and tedious, so we omit the details and only summarize the results from our analysis. We use the value CFL, where $\Delta t = \frac{CFL}{\max |f'(u)|} \Delta x = \frac{CFL}{M^{m-1}} \Delta x$. Clearly, $\lambda = \frac{CFL}{M^{m-1}}$ so in the linear case ($m = 1$), the two are identical. However, in the nonlinear case, the two are different, and while the term λ is more useful for analysis of a given problem, the CFL is useful in comparing different problems and different values of M. For this reason, in this section, we will use both.

Assuming that $\varepsilon \ll M \leq 1$, we can show analytically[1] that the maximal overshoot/undershoot is of the form

$$\alpha_m \times CFL \times M \times \left(\frac{\varepsilon}{M^{2m}}\right)^2, \tag{4}$$

where

$$\alpha_2 = \frac{779}{7560} \qquad \alpha_3 = \frac{6137}{143360} \qquad \alpha_4 = \frac{18871}{637875}.$$

---

[1]Details are omitted for space considerations.

The numerical results in Tables 1 and 2 confirm the formula for the undershoot error of WENO-JS (4) and the values of α_m. Once again, we stress that the above estimates of undershoot error are valid only if ε is small compared to M. When ε is not small enough, we will see different behavior. For this reason, we used the values of ε for which the constants in the overshoot/undershoot values converged. In the tables in this section, the values of ε shown are the largest ones that give the undershoot error accurate up to five digits for listed M.

Remark Note that the exponent 2 outside of the final parentheses in the relation above comes from the fact that we use the power $p = 2$ in the WENO method. In the remainder of this section, we assume that $p = 2$ and we will see this power appears in the oscillations for all methods. Using a different power p in the WENO method results in a corresponding exponent in the dependence of the oscillation on ε and M^{2m}.

3.2 One Step of WENO-JS with Two-Stage Second-Order Runge–Kutta

In this section, we provide an analysis and numerical confirmation for the behavior of the overshoot/undershoot when one time-step is taken using the strong stability preserving two-stage second-order Runge–Kutta (RK2) method [2, 11]

$$u^{(1)} = u^n + \Delta t F(u^n)$$
$$u^{n+1} = u^n + \frac{1}{2}\Delta t F(u^n) + \frac{1}{2}\Delta t F(u^{(1)}).$$

This analysis shows the emerging complexity as higher-order methods are used. As before, we begin with a complete analysis of the linear case and omit the full details.

3.2.1 Linear Case

As before, we assume that $\varepsilon \ll M$. We can show that the maximal overshoot is of the form $P_1(\lambda)\frac{\varepsilon^2}{M^3}$, where P_1 is given by the maximum of two rational functions $P_1(\lambda) = \max\big(p_1(\lambda), p_2(\lambda)\big)$ where

$$p_1 = (14400 - 136800\lambda + 504900\lambda^2 - 860760\lambda^3 + 598228\lambda^4 - 116306\lambda^5$$
$$+ 328413\lambda^6 - 467600\lambda^7 + 270625\lambda^8)/(1244160\lambda^2(4 - 19\lambda + 25\lambda^2)^2)$$
$$p_2 = -(\lambda(2343600 - 19628460\lambda + 75692628\lambda^2 - 177350059\lambda^3 + 276036659\lambda^4$$
$$- 287679978\lambda^5 + 190949760\lambda^6 - 72004875\lambda^7 + 11803125\lambda^8))/(2508800$$
$$(-1 + \lambda)^3(10 - 31\lambda + 25\lambda^2)^2).$$

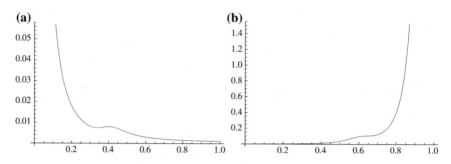

Fig. 4 The functions $p_1(\lambda)$ and $p_2(\lambda)$. (a) $p_1(\lambda)$. (b) $p_2(\lambda)$.

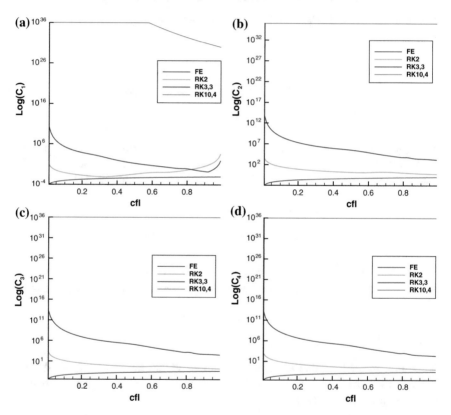

Fig. 5 The coefficient $C_m(CFL)$ of the undershoot of WENO-JS after one time-step. The CFL is between $[0.01, 0.99]$ with an increment of 0.01. (a) WENO-JS, $f(u) = u$. (b) WENO-JS, $f(u) = u^2/2$. (c) WENO-JS, $f(u) = u^3/3$. (d) WENO-JS, $f(u) = u^4/4$.

The undershoot is given by $Q_1(\lambda)\frac{\varepsilon^2}{M^3}$, where $Q(\lambda) = \max\big(q_1(\lambda), q_2(\lambda)\big)$. In this case, $p_1 = q_1$ and $p_2 = q_2$, so the maximal overshoot and undershoot are the same.

Let $C_1 = P_1$ and notice that C_1 is a function of λ that depends on the functions $p_1(\lambda)$ and $p_2(\lambda)$. Using the expressions of P and Q, and the graphs in Figure 4, we can see that when fixing ε small enough and decreasing λ from 1 to 0, the overshoot/undershoot will decrease at first and then increase. This is also reflected in Figure 5(a) in the green line.

3.2.2 Nonlinear Cases

Once again, we must assume that ε is small enough. For our purposes, we require $\varepsilon \ll M^{2m}$. If we take $\Delta t = CFL\Delta x/M^{m-1}$, then after one RK2 time-step, we get the following behavior for the overshoot/undershoot:

$$C_m(CFL)\frac{\varepsilon^2}{M^{4m-1}}, \qquad (5)$$

where the value $C_m(CFL)$ depends only on the CFL number and m and on whether we are looking for an overshoot or undershoot. Figure 5 shows the graphs of C_m as a function of the CFL number for $m = 1, 2, 3, 4$. The numerical results for the undershoot error of WENO-JS with RK2 are given in Table 3, for $CFL = \frac{1}{4}$. These values verify that the undershoot error is proportional to $\frac{\varepsilon^2}{M^{4m-1}}$ and give us the values of $C_m(\frac{1}{4})$ in Table 4.

3.3 One Step of WENO-JS with Higher-Order Time-stepping Schemes

In practice, the explicit strong stability preserving three-stage third-order Runge–Kutta method of Shu and Osher (RK3,3) [2, 11] (given in the Appendix) is frequently used to evolve the solution forward. Another excellent alternative is Ketcheson's explicit strong stability preserving ten-stage fourth-order Runge–Kutta (RK10,4) method [2, 8], given in the Appendix. It is interesting to observe the effect of the higher-order time-stepping method on the overshoots and undershoots produced by WENO-JS. We experimented with these time-stepping methods, and our numerical results for the undershoots with different ε and M suggest that the undershoot error is of the form

$$C_m(CFL)\frac{\varepsilon^2}{M^{4m-1}}, \qquad (6)$$

Table 3 The undershoot error of WENO-JS after one time-step with RK2 using step function initial conditions, $nx = 200$, $CFL = 0.25$.

		$M = 1$	$M = 0.1$	$M = 0.01$	$M = 0.001$	pattern
$f(u) = u$	$\varepsilon = 1e{-}11$	$1.1413e{-}24$	$1.1413e{-}21$	$1.1413e{-}18$	$1.1413e{-}15$	$1.1413 \times 10^{-2}\frac{\varepsilon^2}{M^3}$
	$\varepsilon = 1e{-}12$	$1.1413e{-}26$	$1.1413e{-}23$	$1.1413e{-}20$	$1.1413e{-}17$	
	$\varepsilon = 1e{-}13$	$1.1413e{-}28$	$1.1413e{-}25$	$1.1413e{-}22$	$1.1413e{-}19$	
$f(u) = u^2/2$	$\varepsilon = 1e{-}20$	$5.4665e{-}40$	$5.4665e{-}33$	$5.4665e{-}26$	$5.4665e{-}19$	$5.4665\frac{\varepsilon^2}{M^7}$
	$\varepsilon = 1e{-}21$	$5.4665e{-}42$	$5.4665e{-}35$	$5.4665e{-}28$	$5.4665e{-}21$	
	$\varepsilon = 1e{-}22$	$5.4665e{-}44$	$5.4665e{-}37$	$5.4665e{-}30$	$5.4665e{-}23$	
$f(u) = u^3/3$	$\varepsilon = 1e{-}26$	$2.0345e{-}52$	$2.0345e{-}41$	$2.0345e{-}30$	$2.0345e{-}19$	$2.0345\frac{\varepsilon^2}{M^{11}}$
	$\varepsilon = 1e{-}27$	$2.0345e{-}54$	$2.0345e{-}43$	$2.0345e{-}32$	$2.0345e{-}21$	
	$\varepsilon = 1e{-}28$	$2.0345e{-}56$	$2.0345e{-}45$	$2.0345e{-}34$	$2.0345e{-}23$	
$f(u) = u^4/4$	$\varepsilon = 1e{-}31$	$1.4155e{-}62$	$1.4155e{-}47$	$1.4155e{-}32$	$1.4155e{-}17$	$1.4155\frac{\varepsilon^2}{M^{15}}$
	$\varepsilon = 1e{-}32$	$1.4155e{-}64$	$1.4155e{-}49$	$1.4155e{-}34$	$1.4155e{-}19$	
	$\varepsilon = 1e{-}33$	$1.4155e{-}66$	$1.4155e{-}51$	$1.4155e{-}36$	$1.4155e{-}21$	

Table 4 The undershoot error of the WENO method after one time-step of the time-stepping method. Initial condition is the using step function initial conditions. The spatial number of points is $nx = 200$, with $CFL = 0.25$.

Time	Space	$C_1(0.25)$	$C_2(0.25)$	$C_3(0.25)$	$C_4(0.25)$
FE	WENO-JS	2.3354×10^{-3}	2.5761×10^{-2}	1.0702×10^{-2}	7.3960×10^{-2}
	WENO-M	6.9447×10^{-3}	1.1345×10^{-1}	4.6954×10^{-2}	3.2318×10^{-2}
	WENO-Z	3.0301×10^{-3}	3.4021×10^{-2}	1.4021×10^{-2}	9.6069×10^{-3}
RK2	WENO-JS	1.1413×10^{-2}	5.4665	2.0345	1.4155
	WENO-M	3.9962×10^{-2}	1.7779×10^{1}	6.4421	4.4725
	WENO-Z	1.3381×10^{-2}	5.0724	1.8059	1.2517
RK3, 3	WENO-JS	2.6639×10^{3}	6.6192×10^{6}	2.5725×10^{6}	1.7984×10^{6}
	WENO-M	2.4487×10^{4}	2.4851×10^{7}	9.8591×10^{6}	6.9289×10^{6}
	WENO-Z	3.2231×10^{3}	5.9477×10^{6}	2.3146×10^{6}	1.6191×10^{6}
RK10, 4	WENO-JS	2.9481×10^{45}	2.0732×10^{55}	8.6113×10^{54}	6.0468×10^{54}
	WENO-M	1.0722×10^{46}	6.5880×10^{55}	2.7364×10^{55}	1.9215×10^{55}
	WENO-Z	3.2443×10^{45}	1.8348×10^{55}	7.6210×10^{54}	5.3515×10^{54}

where C_m depends on the CFL number and m. However, the values of the coefficient C_m increase with the order of the time-stepping method, as shown in Figure 5. The values of $C_m(\frac{1}{4})$ for both RK3,3 and RK10,4 are given in Table 4. Clearly, the coefficients for the RK10,4 are much larger, indicating larger oscillations. Of course, higher-order time-stepping methods allow us to use larger time-steps and therefore larger values of CFL, for both accuracy and stability. Nevertheless, the sensitivity of the overshoot/undershoot to the order of the time-stepping method exceeds this benefit.

(a)   **(b)**

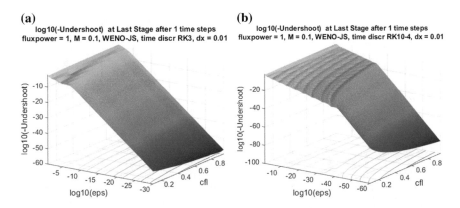

Fig. 6 Undershoot of WENO-JS after one time-step. $f(u) = u$, $M = 0.1$. Left: RK3,3, right: RK10,4. (a) WENO-JS with RK3,3, ε=1e-1 to 1e-30. (b) WENO-JS with RK10,4, ε=1e-1 to 1e-30.

Note that we have not determined exactly how ε_c, the critical value of ε at which the undershoot error starts the asymptotic behavior, depends on M, CFL number, time schemes, and the power m in $f(u)$. However, numerical tests show that higher-order time-stepping methods require smaller ε to reach the asymptotic region. In Figure 6, we consider the linear case $f(u) = u$ using $M = 0.1$. We see that RK10,4 requires much smaller ε than RK3,3 to reach the asymptotic region.

3.4 Other WENO Methods

We ran the same numerical tests using the WENO-M and WENO-Z methods. We observed that WENO-M and WENO-Z have a similar behavior to WENO-JS, in the sense that the undershoot error is of the form

$$C_m(CFL) \times M \times \left(\frac{\varepsilon}{M^{2m}}\right)^2 .$$

The difference between methods results in different values of C_m. The values of C_m as a function of the CFL number for WENO-M and WENO-Z are given in Figure 7. In these Figures, the CFL number is between $[0.01, 0.99]$ with an increment 0.01. Again, we see that the higher-order the time scheme we use, the larger the coefficient C_m we have for the undershoot error, but the different WENO methods have similar coefficients. A comparison of the values of the $C_m(0.25)$ is given in Table 4, where we observe that the values are larger for WENO-M, but generally of the same order.

To study this behavior on a smoother function, with only one discontinuity, we consider the case where the initial condition u_0 is not a step-function:

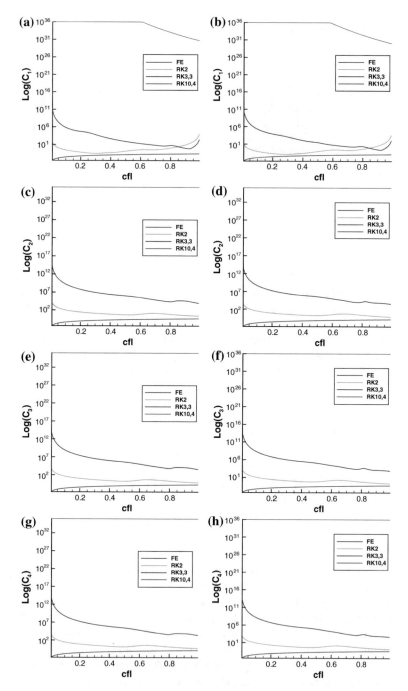

Fig. 7 The coefficient $C_m(CFL)$ of the undershoot of WENO-M and WENO-Z after one time-step. (a) WENO-M, $f(u) = u$. (b) WENO-Z, $f(u) = u$. (c) WENO-M, $f(u) = u^2/2$. (d) WENO-Z, $f(u) = u^2/2$. (e) WENO-M, $f(u) = u^3/3$. (f) WENO-Z, $f(u) = u^3/3$. (g) WENO-M, $f(u) = u^4/4$. (h) WENO-Z, $f(u) = u^4/4$.

Table 5 The undershoot error of the WENO method after one time-step of the time-stepping method with the nonstep function initial condition (7). The spatial number of points is $nx = 200$, with $CFL = 0.25$.

Time	Space	$C_1(0.25)$	$C_2(0.25)$	$C_3(0.25)$	$C_4(0.25)$
FE	WENO-JS	4.1831×10^{12}	4.8296×10^{12}	4.8295×10^{12}	4.8295×10^{12}
	WENO-M	1.1561×10^{13}	1.5339×10^{13}	1.5338×10^{13}	1.5338×10^{13}
	WENO-Z	4.3273×10^{12}	4.2740×10^{12}	4.2739×10^{12}	4.2739×10^{12}
RK2	WENO-JS	1.4404×10^{13}	3.3623×10^{16}	3.3866×10^{16}	3.4110×10^{16}
	WENO-M	3.9712×10^{13}	1.0678×10^{17}	2.0755×10^{17}	1.0833×10^{17}
	WENO-Z	1.4864×10^{13}	2.9755×10^{16}	2.9970×10^{16}	3.0186×10^{16}
RK3, 3	WENO-JS	7.8146×10^{12}	3.9911×10^{22}	4.0461×10^{22}	4.1034×10^{22}
	WENO-M	2.1346×10^{13}	1.2676×10^{23}	1.2850×10^{23}	1.3032×10^{23}
	WENO-Z	7.9869×10^{12}	3.5320×10^{22}	3.5806×10^{22}	3.6313×10^{22}
RK10, 4	WENO-JS	2.9505×10^{45}	1.0158×10^{71}	1.5943×10^{71}	1.6888×10^{71}
	WENO-M	1.0731×10^{46}	4.7812×10^{71}	5.0610×10^{71}	5.3607×10^{71}
	WENO-Z	3.2469×10^{45}	1.3328×10^{71}	1.4108×10^{71}	1.4945×10^{71}

$$u_0(x) = \begin{cases} 0, & x \in [-1, -1/8] \\ Me \cdot e^{\frac{1}{16(x-1/8)^2-1}}, & x \in (-1/8, 1/8] \\ 0, & x \in (1/8, 1] \end{cases} \tag{7}$$

We refer to this as the "shark" function due to its appearance which resembles a shark fin. Our numerical results show that the undershoots of WENO methods have similar forms to those in the previous subsection where we use step function initial conditions. The values of C_m shown in Table 5 are much larger than for the step function initial conditions, and the ε needed for the behavior to settle is much smaller, but the qualitative behavior is the same.

4 The Effect of the Sensitivity Parameter Over Time Integration

In the previous section, we studied the effect of the value of ε and the size of the jump M on one step. In this section, we discuss how the undershoots and overshoots propagate over time and their dependence on both ε and M. We observe that the behavior of oscillations for all three WENO methods, WENO-JS, WENO-M, and WENO-Z, is similar and that the choice of time integration also does not affect the long time behavior of the oscillations.

We ran the examples in Section 4 up to time $T = 2$, for WENO-JS, WENO-M, and WENO-Z, with time-stepping methods RK3,3 and RK10,4. Figure 8 shows that for $\varepsilon = 1.e - 6$, the undershoot varies over time integration. In all cases, the

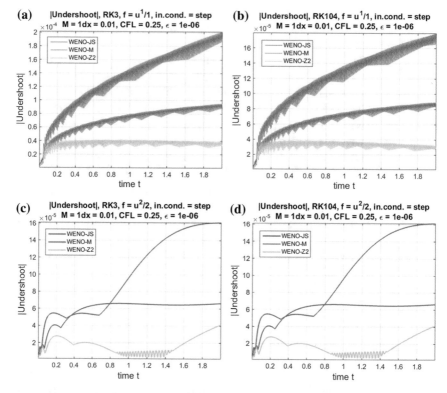

Fig. 8 Behavior of the undershoots for different WENO methods. (a) WENO with RK3,3, $m = 1$. (b) WENO with RK10,4, $m = 1$. (c) WENO with RK3,3, $m = 2$. (d) WENO with RK10,4, $m = 2$.

undershoot is larger for RK3,3 than for RK10,4, and the undershoot for WENO-M is the largest, followed by WENO-JS, and WENO-Z has the smallest undershoots. However, we also see that these overshoots are all of the same order of magnitude: approximately 10^{-5} for $\varepsilon = 1.e - 6$. Numerous experiments with different values of ε, shown in Figure 9, confirm that the value of ε is the major determinant of the size of the oscillation. The major observation from these figures is that the long time behavior of the oscillations scales with $\sqrt{\varepsilon}$. This pattern is confirmed and further strengthened in our study of the maximal undershoot observed over 800 time-steps. The clear pattern that emerges from the numerical results in Tables 6 and 7 is that the undershoot is

$$\mathcal{O}(\sqrt{\varepsilon}/M^{m-1}). \tag{8}$$

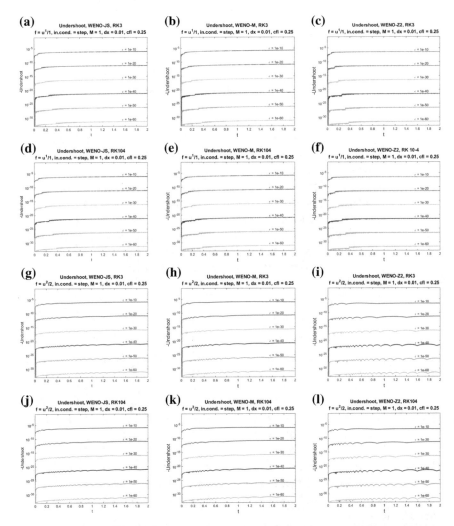

Fig. 9 Undershoot for WENO-JS with unit step initial condition, $CFL = 0.25$. Figures (a)–(f) have $m = 1$, while Figures (g)–(l) have $m = 2$. (a) WENO-JS with RK3,3. (b) WENO-M with RK3,3. (c) WENO-Z with RK3,3. (d) WENO-JS with RK10,4. (e) WENO-M with RK10,4. (f) WENO-Z with RK10,4. (g) WENO-JS with RK3,3. (h) WENO-M with RK3,3. (i) WENO-Z with RK3,3. (j) WENO-JS with RK10,4. (k) WENO-M with RK10,4. (l) WENO-Z with RK10,4.

Table 6 Maximum undershoot values for different WENO methods over 800 time-steps with RK3,3. $nx = 200$, $CFL = 0.25$.

WENO-JS with RK3,3

		$M = 1$	$M = 0.1$	$M = 0.01$	$M = 0.001$	pattern
$f(u) = u$	$\varepsilon = 1e\text{-}12$	9.3143E-08	9.4454E-08	9.6084E-08	9.5517E-08	$0.09\,\sqrt{\varepsilon}$
	$\varepsilon = 1e\text{-}14$	9.1005E-09	9.3143E-09	9.4454E-09	9.6084E-09	
	$\varepsilon = 1e\text{-}16$	8.8992E-10	9.1005E-10	9.3143E-10	9.4454E-10	
$f(u) = u^2/2$	$\varepsilon = 1e\text{-}22$	2.0375E-12	2.1317E-11	1.9802E-10	1.3910E-09	$0.2\,\sqrt{\varepsilon}/M$
	$\varepsilon = 1e\text{-}24$	2.0936E-13	2.0530E-12	2.0619E-11	1.9183E-10	
	$\varepsilon = 1e\text{-}26$	1.8374E-14	2.0375E-13	2.1317E-12	1.9802E-11	
$f(u) = u^3/3$	$\varepsilon = 1e\text{-}32$	3.8517E-17	4.3820E-15	4.2972E-13	4.3048E-11	$0.4\,\sqrt{\varepsilon}/M^2$
	$\varepsilon = 1e\text{-}34$	4.0889E-18	4.6507E-16	4.8072E-14	4.3279E-12	
	$\varepsilon = 1e\text{-}36$	4.2655E-19	4.2884E-17	4.2528E-15	4.6742E-13	
$f(u) = u^4/4$	$\varepsilon = 1e\text{-}42$	7.7903E-22	7.1192E-19	7.1688E-16	8.3674E-13	$0.8\,\sqrt{\varepsilon}/M^3$
	$\varepsilon = 1e\text{-}44$	7.5143E-23	7.3278E-20	8.2754E-17	7.4515E-14	
	$\varepsilon = 1e\text{-}46$	7.2119E-24	7.7690E-21	8.6189E-18	8.4379E-15	

WENO-JS with RK3,3

		$M = 1$	$M = 0.1$	$M = 0.01$	$M = 0.001$	pattern
$f(u) = u$	$\varepsilon = 1e\text{-}12$	1.9108E-07	1.9576E-07	1.9905E-07	2.0141E-07	$0.2\,\sqrt{\varepsilon}$
	$\varepsilon = 1e\text{-}14$	1.8409E-08	1.9108E-08	1.9576E-08	1.9905E-08	
	$\varepsilon = 1e\text{-}16$	1.8016E-09	1.8409E-09	1.9108E-09	1.9576E-09	
$f(u) = u^2/2$	$\varepsilon = 1e\text{-}22$	3.4297E-12	3.2013E-11	2.1447E-10	1.2370E-09	$0.3\,\sqrt{\varepsilon}/M$
	$\varepsilon = 1e\text{-}24$	3.3031E-13	3.4222E-12	2.7275E-11	1.8059E-10	
	$\varepsilon = 1e\text{-}26$	3.1131E-14	3.4297E-13	3.2013E-12	2.1447E-11	
$f(u) = u^3/3$	$\varepsilon = 1e\text{-}32$	6.2104E-17	5.9566E-15	7.1510E-13	5.0514E-11	$0.6\,\sqrt{\varepsilon}/M^2$
	$\varepsilon = 1e\text{-}34$	6.1962E-18	5.9036E-16	6.8629E-14	6.4428E-12	
	$\varepsilon = 1e\text{-}36$	6.0855E-19	6.1169E-17	6.4301E-15	7.1007E-13	
$f(u) = u^4/4$	$\varepsilon = 1e\text{-}42$	1.2019E-21	1.1666E-18	1.2908E-15	1.1876E-12	$1.2\,\sqrt{\varepsilon}/M^3$
	$\varepsilon = 1e\text{-}44$	1.2258E-22	1.1601E-19	1.2513E-16	1.2989E-13	
	$\varepsilon = 1e\text{-}46$	1.2459E-23	1.1654E-20	1.2143E-17	1.3348E-14	

WENO-Z with RK3,3

		$M = 1$	$M = 0.1$	$M = 0.01$	$M = 0.001$	pattern
$f(u) = u$	$\varepsilon = 1e\text{-}12$	9.3141E-08	9.4217E-08	9.6086E-08	9.5518E-08	$0.09\,\sqrt{\varepsilon}$
	$\varepsilon = 1e\text{-}14$	9.0855E-09	9.3146E-09	9.4450E-09	9.6084E-09	
	$\varepsilon = 1e\text{-}16$	8.8874E-10	9.1067E-10	9.3144E-10	9.4454E-10	
$f(u) = u^2/2$	$\varepsilon = 1e\text{-}22$	2.0086E-12	2.1317E-11	1.9802E-10	1.3910E-09	$0.2\,\sqrt{\varepsilon}/M$
	$\varepsilon = 1e\text{-}24$	2.1053E-13	2.0530E-12	2.0619E-11	1.9183E-10	
	$\varepsilon = 1e\text{-}26$	1.8560E-14	2.0375E-13	2.1317E-12	1.9802E-11	
$f(u) = u^3/3$	$\varepsilon = 1e\text{-}32$	3.9087E-17	4.3820E-15	4.2972E-13	4.3048E-11	$0.4\,\sqrt{\varepsilon}/M^2$
	$\varepsilon = 1e\text{-}34$	3.9973E-18	4.6507E-16	4.8072E-14	4.3279E-12	
	$\varepsilon = 1e\text{-}36$	4.2212E-19	4.2884E-17	4.2528E-15	4.6742E-13	
$f(u) = u^4/4$	$\varepsilon = 1e\text{-}42$	7.7883E-22	7.1192E-19	7.1688E-16	8.3674E-13	$0.8\,\sqrt{\varepsilon}/M^3$
	$\varepsilon = 1e\text{-}44$	7.5875E-23	7.3278E-20	8.2754E-17	7.4515E-14	
	$\varepsilon = 1e\text{-}46$	7.3054E-24	7.7690E-21	8.6189E-18	8.4379E-15	

Table 7 Maximum undershoot values for different WENO methods over 800 time-steps with RK10,4. $nx = 200$, $cfl = 0.25$.

WENO-JS with RK10,4

		$M = 1$	$M = 0.1$	$M = 0.01$	$M = 0.001$	pattern
$f(u) = u$	$\varepsilon = $ 1e-12	8.7682E-08	8.9456E-08	9.0061E-08	8.9096E-08	$0.09\,\sqrt{\varepsilon}$
	$\varepsilon = $ 1e-14	8.5621E-09	8.7682E-09	8.9456E-09	9.0061E-09	
	$\varepsilon = $ 1e-16	8.3900E-10	8.5621E-10	8.7682E-10	8.9456E-10	
$f(u) = u^2/2$	$\varepsilon = $ 1e-22	2.0375E-12	2.1317E-11	1.9801E-10	1.3908E-09	$0.2\,\sqrt{\varepsilon}/M$
	$\varepsilon = $ 1e-24	2.0938E-13	2.0533E-12	2.0622E-11	1.9184E-10	
	$\varepsilon = $ 1e-26	1.8376E-14	2.0375E-13	2.1317E-12	1.9801E-11	
$f(u) = u^3/3$	$\varepsilon = $ 1e-32	3.8523E-17	4.3814E-15	4.2966E-13	4.3052E-11	$0.4\,\sqrt{\varepsilon}/M^2$
	$\varepsilon = $ 1e-34	4.0884E-18	4.6508E-16	4.8073E-14	4.3274E-12	
	$\varepsilon = $ 1e-36	4.2653E-19	4.2891E-17	4.2537E-15	4.6747E-13	
$f(u) = u^4/4$	$\varepsilon = $ 1e-42	7.7905E-22	7.1203E-19	7.1704E-16	8.3673E-13	$0.8\,\sqrt{\varepsilon}/M^3$
	$\varepsilon = $ 1e-44	7.5150E-23	7.3265E-20	8.2744E-17	7.4536E-14	
	$\varepsilon = $ 1e-46	7.2126E-24	7.7681E-21	8.6190E-18	8.4371E-15	

WENO-M with RK10,4

		$M = 1$	$M = 0.1$	$M = 0.01$	$M = 0.001$	pattern
$f(u) = u$	$\varepsilon = $ 1e-12	1.7391E-07	1.7691E-07	1.7946E-07	1.7970E-07	$0.17\,\sqrt{\varepsilon}$
	$\varepsilon = $ 1e-14	1.7005E-08	1.7391E-08	1.7691E-08	1.7946E-08	
	$\varepsilon = $ 1e-16	1.6484E-09	1.7005E-09	1.7391E-09	1.7691E-09	
$f(u) = u^2/2$	$\varepsilon = $ 1e-22	3.4304E-12	3.2001E-11	2.1465E-10	1.2365E-09	$0.3\,\sqrt{\varepsilon}/M$
	$\varepsilon = $ 1e-24	3.3045E-13	3.4220E-12	2.7258E-11	1.8063E-10	
	$\varepsilon = $ 1e-26	3.1149E-14	3.4304E-13	3.2001E-12	2.1465E-11	
$f(u) = u^3/3$	$\varepsilon = $ 1e-32	6.2056E-17	5.9619E-15	7.1537E-13	5.0485E-11	$0.6\,\sqrt{\varepsilon}/M^2$
	$\varepsilon = $ 1e-34	6.1915E-18	5.8986E-16	6.8670E-14	6.4411E-12	
	$\varepsilon = $ 1e-36	6.0807E-19	6.1120E-17	6.4351E-15	7.1012E-13	
$f(u) = u^4/4$	$\varepsilon = $ 1e-42	1.2024E-21	1.1672E-18	1.2913E-15	1.1874E-12	$1.2\,\sqrt{\varepsilon}/M^3$
	$\varepsilon = $ 1e-44	1.2262E-22	1.1607E-19	1.2518E-16	1.2989E-13	
	$\varepsilon = $ 1e-46	1.2462E-23	1.1660E-20	1.2149E-17	1.3350E-14	

WENO-Z with RK10,4

		$M = 1$	$M = 0.1$	$M = 0.01$	$M = 0.001$	pattern
$f(u) = u$	$\varepsilon = $ 1e-12	8.7426E-08	8.9384E-08	9.0065E-08	8.9096E-08	$0.09\,\sqrt{\varepsilon}$
	$\varepsilon = $ 1e-14	8.5971E-09	8.7851E-09	8.9456E-09	9.0061E-09	
	$\varepsilon = $ 1e-16	8.4214E-10	8.5788E-10	8.7687E-10	8.9456E-10	
$f(u) = u^2/2$	$\varepsilon = $ 1e-22	2.0084E-12	2.1316E-11	1.9801E-10	1.3908E-09	$0.2\,\sqrt{\varepsilon}/M$
	$\varepsilon = $ 1e-24	2.1055E-13	2.0533E-12	2.0622E-11	1.9184E-10	
	$\varepsilon = $ 1e-26	1.8562E-14	2.0375E-13	2.1317E-12	1.9801E-11	
$f(u) = u^3/3$	$\varepsilon = $ 1e-32	3.9095E-17	4.3814E-15	4.2966E-13	4.3052E-11	$0.4\,\sqrt{\varepsilon}/M^2$
	$\varepsilon = $ 1e-34	3.9963E-18	4.6508E-16	4.8073E-14	4.3274E-12	
	$\varepsilon = $ 1e-36	4.2208E-19	4.2891E-17	4.2537E-15	4.6747E-13	
$f(u) = u^4/4$	$\varepsilon = $ 1e-42	7.7883E-22	7.1203E-19	7.1704E-16	8.3673E-13	$0.8\,\sqrt{\varepsilon}/M^3$
	$\varepsilon = $ 1e-44	7.5877E-23	7.3265E-20	8.2744E-17	7.4536E-14	
	$\varepsilon = $ 1e-46	7.3057E-24	7.7681E-21	8.6190E-18	8.4371E-15	

Table 8 Critical CFL stability limits for the linear and nonlinear equations with $\varepsilon = 1e - 50$.

Time-stepping method	WENO scheme	Linear Equation		Burgers' Equation	
		CFL_{shark}	CFL_{step}	CFL_{shark}	CFL_{step}
RK3,3	JS	0.82	0.78	1.48	1.44
	M	0.79	0.73	1.66	1.48
	Z	0.78	0.73	1.58	1.44
RK10,4	JS	3.15	3.20	6.64	5.37
	M	2.91	2.94	6.96	5.79
	Z	3.07	3.03	6.98	5.82

4.1 Stability Limits

The long time behavior of the solution led us to investigate the stability limits of the numerical methods used. We find that nonlinear instability occurs at certain CFL numbers that depend most strongly on the time-stepping algorithm and to some extent on the WENO method. Small variations depending on the initial condition are observed as well when we tested the initial functions (2) and (7). When ε is reasonably small, its precise value does not seem to play a significant role. The largest stable value of CFL for the different methods is given in Table 8.

In the linear case, with a given reasonably small ε, the differences between the allowable time-step for the different WENO methods are not significant. However, the choice of time-stepping method makes a difference. The CFL for the ten-stage fourth-order RK10,4 is significantly larger than that of three-stage third-order RK3,3. Even when we take into account that RK10,4 requires ten stages and RK3,3 only three, we can conclude that RK10,4 has an advantage, because it allows for higher *effective* CFL numbers when taking into account the number of stages. For example, for the shark profile RK3,3 with WENO-Z requires at most 0.78 CFL ($0.78/3 = 0.26$), while RK10,4 for the same WENO method has a limitation of 3.07 CFL ($3.07/10 = 0.307$).

The linear and nonlinear cases exhibit remarkably different behaviors in terms of the allowable CFL, but the nonlinear cases all behaved similarly for $m = 2, 3, 4$. For this reason, the results for Burgers' equation serve as a representative for the nonlinear problems. For Burgers' equation, we observed that WENO-JS had the smallest allowable time-step, but that all WENO methods behaved generally the same. The time-stepping methods result in very different stability limits, and once again, we observe that RK10,4 has a significant advantage over RK3,3 even when normalizing for the number of function evaluations.

Acknowledgments This project is part of an IMA-funded WhAM! workshop and ongoing program. We thank the IMA for its support. Our thanks to Daniel Higgs who produced the graphs in Section 3.

Appendix A

Strong Stability Preserving Runge–Kutta Time Evolution Methods

To preserve the designed nonlinear stability properties of the WENO methods, it is advisable to use strong stability preserving Runge–Kutta methods [2]. In this work, we use

The three-stage third order SSP method by Shu and Osher [11, 12] (RK3,3)

$$u^{(1)} = u^n + \Delta t F(u^n)$$
$$u^{(2)} = \frac{3}{4}u^n + \frac{1}{4}u^{(1)} + \frac{1}{4}\Delta t F(u^{(1)})$$
$$u^{n+1} = \frac{1}{3}u^n + \frac{2}{3}u^{(2)} + \frac{2}{3}\Delta t F(u^{(2)})$$

The ten-stage fourth order SSP method by Ketcheson [8] (RK10,4)

$$u^{(1)} = u^n + \frac{1}{6}\Delta t F(u^n)$$
$$u^{(k+1)} = u^{(k)} + \frac{1}{6}\Delta t F(u^{(k)}) \quad \text{for } k = 1, 2, 3$$
$$u^{(5)} = \frac{3}{5}u^n + \frac{2}{5}u^{(4)} + \frac{1}{15}\Delta t F(u^{(4)})$$
$$u^{(k+1)} = u^{(k)} + \frac{1}{6}\Delta t F(u^{(k)}) \quad \text{for } k = 5, 6, 7, 8$$
$$u^{n+1} = \frac{1}{25}u^n + \frac{9}{25}u^{(4)} + \frac{3}{5}u^{(9)} + \frac{3}{50}\Delta t F(u^{(4)}) + \frac{1}{10}\Delta t F(u^{(9)}).$$

References

1. R. BORGES, M. CARMONA, B. COSTA, AND W. S. DON, *An improved weighted essentially non-oscillatory scheme for hyperbolic conservation laws*, Journal of Computational Physics, 227 (2008), pp. 3191 – 3211.
2. S. GOTTLIEB, D. I. KETCHESON, AND C.- W. SHU, *Strong Stability Preserving Runge–Kutta and Multistep Time Discretizations*, World Scientific Press, 2011.

3. A. HARTEN, B. ENGQUIST, S. OSHER AND S. CHAKRAVARTHY, *Uniformly High Order Essentially Non-Oscillatory Schemes I*, SIAM Journal on Numerical Analysis, **vol. 24**, (1987), pp.279-309.

4. A. HARTEN, B. ENGQUIST, S. OSHER AND S. CHAKRAVARTHY, *Uniformly High Order Essentially Non-Oscillatory Schemes III*, Journal of Computational Physics, **vol. 71**, (1987), pp.231-303.

5. A. K. HENRICK, T. D. ASLAM, AND J. M. POWERS, *Mapped weighted essentially non-oscillatory schemes: Achieving optimal order near critical points*, Journal of Computational Physics, 207 (2005), pp. 542 – 567.

6. G. B. JACOBS AND W.- S. DON, *A high-order WENO-Z finite difference based particle-source-in-cell method for computation of particle-laden flows with shocks*, Journal of Computational Physics, 228 (2009), pp. 1365 – 1379.

7. G.- S. JIANG AND C.- W. SHU, *Efficient implementation of weighted ENO schemes*, Journal of Computational Physics, 126 (1996), pp. 202 – 228.

8. D. I. KETCHESON, *Highly efficient strong stability preserving Runge–Kutta methods with low-storage implementations*, SIAM Journal on Scientific Computing, 30 (2008), pp. 2113–2136.

9. R.J. LEVEQUE, *Numerical Methods for Conservation Laws*, (Lectures in Mathematics), Birkhauser, Basel; Boston; Berlin; 1992.

10. X.- D. LIU, S. OSHER AND T. CHAN, *Weighted Essentially Non-Oscillatory Schemes*, Journal of Computational Physics, **v. 115**, (1994), p.200.

11. C.- W. SHU, *Total-Variation-Diminishing Time Discretizations*, SIAM J. Sci. Stat. Comput. **v. 9**, (1988), pp.1073-1084.

12. C.- W. SHU AND S. OSHER, *Efficient Implementation of Essentially Non-Oscillatory Shock-Capturing Schemes*, J. Comput. Phys. **v. 77**, (1988), pp.439-471.

13. C.- W. SHU, *High order finite difference and finite volume WENO schemes and discontinuous Galerkin methods for CFD*, International Journal of Computational Fluid Dynamics, v17 (2003), pp.107-118.

Study of a Mixed Dispersal Population Dynamics Model

Marina Chugunova, Baasansuren Jadamba, Chiu-Yen Kao,
Christine Klymko, Evelyn Thomas and Bingyu Zhao

Abstract In this paper, we consider a mixed dispersal model with periodic and Dirichlet boundary conditions and its corresponding linear eigenvalue problem. This model describes the time evolution of a population which disperses both locally and nonlocally. We investigate how long time dynamics depend on the parameter values. Furthermore, we study the minimization of the principal eigenvalue under the constraints that the resource function is bounded from above and below, and with a fixed total integral. Biologically, this minimization problem is motivated by the question of determining the optimal spatial arrangement of favorable and unfavorable regions for the species to die out more slowly or survive more easily. Our numerical simulations indicate that the optimal favorable region tends to be a simply connected

M. Chugunova
Department of Mathematics, Claremont Graduate University, Claremont,
CA 91711, USA
e-mail: marina.chugunova@cgu.edu

B. Jadamba
School of Mathematical Sciences, Rochester Institute of Technology,
Rochester, NY 14623, USA
e-mail: bxjsma@rit.edu

C.-Y. Kao (✉)
Department of Mathematical Sciences, Claremont McKenna College,
Claremont, CA 91711, USA
e-mail: ckao@cmc.edu

C. Klymko
Center for Applied Scientific Computing, Lawrence Livermore National
Laboratory, Livermore, CA 94550, USA
e-mail: klymko1@llnl.gov

E. Thomas
Department of Mathematics and Statistics, University of Maryland
Baltimore County, Baltimore, MD 21250, USA
e-mail: ekthomas@umbc.edu

B. Zhao
Department of Applied Mathematics, Brown University,
Providence, RI 02912, USA
e-mail: bingyu_zhao@brown.edu

© Springer Science+Business Media New York 2016   51
S.C. Brenner (ed.), *Topics in Numerical Partial Differential Equations
and Scientific Computing*, The IMA Volumes in Mathematics
and its Applications 160, DOI 10.1007/978-1-4939-6399-7_3

domain. Numerous results are shown to demonstrate various scenarios of optimal favorable regions for periodic and Dirichlet boundary conditions.

1 Introduction

To describe the dispersal of species, population dynamics models are commonly used. Random dispersal describes the movement of organisms between adjacent spatial locations [6, 30]. However, the movement of some organisms such as seeds of plants can occur between nonadjacent spatial locations and is thus nonlocal. In the recent years, there have been extensive studies on nonlocal models [1, 2, 7, 10–17, 21, 22, 27–29, 31] and mixed models [18, 24, 32]. Here, we work on a mixed model which was proposed in [24] and study the time evolution problem and optimization of its corresponding eigenvalue problem under a heterogeneous environment.

The model which describes the species adopting both local and nonlocal dispersal is of the form

$$\frac{\partial u}{\partial t} = d\left[\tau \Delta u + (1 - \tau)\, \mathcal{K} u\right] + f(x, u), \quad t > 0, x \in \mathbb{R}^N \tag{1}$$

where $u(x, t)$ denotes the density of species at location x and time t, and the expression $\Delta = \Sigma_{i=1}^{N} \frac{\partial^2}{\partial x_i^2}$ is the Laplace operator in $\mathbb{R}^N$ accounting for random dispersal of species. The nonlocal operator $\mathcal{K}$ is defined by

$$(\mathcal{K} u)(x) := \int_{\mathbb{R}^N} k\left(|x - y|\right) u(y) dy - u(x) \tag{2}$$

where $k = k(r)$ is a smooth, monotone decreasing function with compact support and $k(r)$ satisfies

$$\omega_N \int_0^\infty k(r) r^{N-1} dr = 1 \tag{3}$$

where ω_N denotes the area of the surface of the N-dimensional unit ball. Additionally, d is a positive constant which measures the total number of dispersal individuals per unit time, the constant $0 < \tau \leq 1$ measures the fraction of individuals adopting random dispersal, and, if the logistic model is used, the reaction term is

$$f(x, u) = u(m(x) - u) \tag{4}$$

where $m(x)$ represents the resource function. In the biological applications, it is common to assume that $m(x)$ satisfies

$$-m_1 \leq m(x) \leq m_2 \text{ and } \int_\Omega m(x) dx = M \tag{5}$$

where m_1 and m_2 are positive constants and the constant M could be positive or negative depending on whether the environment is friendly or hostile. This resource function $m(x)$ is positive in the favorable part of habitat and negative in unfavorable one. For the kernel function in (3), we use $k(r) = k^*(r/\delta)/\delta^N$ with

$$
k^*(r) = \begin{cases} C_N \exp\left(\frac{1}{|r|^2-1}\right) & \text{for } |r| < 1 \\ 0 & \text{for } |r| \geq 1 \end{cases}
\tag{6}
$$

where C_N is chosen so that (3) is satisfied. $C_N \approx 2.2523$ in one dimension while $C_N \approx 2.1436$ in two dimensions.

In this paper, we consider the mixed dispersal model (1) on a bounded domain Ω with Dirichlet and periodic boundary conditions. The model (1) on Ω with Dirichlet boundary conditions is given by

$$
\begin{cases} u_t = d\left[\tau \Delta u + (1-\tau)\mathcal{K}u\right] + u(m(x) - u) & \text{for } x \in \Omega, \\ u(x, t) = 0 & \text{for } x \in \partial\Omega. \end{cases}
\tag{7}
$$

One can view this model as the model on $\mathbb{R}^N$ by a zero extension of $u(x, t)$ from Ω to $\mathbb{R}^N \backslash \Omega$.

The model (1) with periodic boundary conditions is given by

$$
\begin{cases} u_t = d\left[\tau \Delta u + (1-\tau)\mathcal{K}u\right] + u(m(x) - u) & \text{for } x \in \mathbb{R}^N, \\ u(x, t) = u(x + p, t) & \text{for } x \in \mathbb{R}^N, \end{cases}
\tag{8}
$$

where $p = (p_1, p_2, \ldots, p_N)$ is a constant vector and the condition $u(x) = u(x + p)$ for $x \in \mathbb{R}^N$ is the so-called p-periodic function. One can view this as a periodic extension from a finite domain $\Omega = (0, p_1) \times (0, p_2) \times \cdots \times (0, p_N)$ to $\mathbb{R}^N$. When the periodic conditions are considered, $m(x)$ is assumed to be p-periodic as well.

In early publications, the focus was on the local dispersal corresponding to $\tau = 1$ in the model (7). The long-term dynamics were analyzed in terms of values of the diffusion parameter d on a bounded domain $\Omega \subset \mathbb{R}^N$ with Dirichlet boundary conditions [4, 5]. It was found that whether $u = 0$ is a global attractor depends on the relationship of d and the principal positive eigenvalue Λ_1 (the smallest positive simple eigenvalue with a corresponding eigenfunction with no sign change) of the indefinite weight eigenvalue problem

$$
-\Delta\psi = \Lambda m(x)\psi
$$

where $m(x)$ satisfies $\text{mes}\{\Omega^+ : m(x) > 0\} \neq 0$ (the favorable region), $\text{mes}\{\Omega^- : m(x) < 0\} \neq 0$ (the unfavorable region), and $\int m(x) = M < 0$. Here, $\text{mes}\{X\}$ denotes the Lebesgue measure of X. It was also shown in [4, 5] that the model (7) with $\tau = 1$ yields a unique positive steady state which is a global attractor for nonnegative nontrivial solutions, provided d is sufficiently small, namely $d < 1/\Lambda_1$.

On the contrary, the solution $u = 0$ is a global attractor for nonnegative solutions if $d > 1/\Lambda_1$ so that the population tends to extinction.

This motivated studies on the minimization of the principal positive eigenvalue in terms of spatial heterogeneity of the resource $m(x)$, which allows species with larger dispersal rate d to survive. In [20, 25], the minimization of the principal eigenvalue was studied with Dirichlet, Neumann, and Robin boundary conditions. The optimal arrangement of $m(x)$ for Dirichlet boundary conditions prefers Ω^+ be a simply connected region away from the boundary on the convex domain. However, for Neumann boundary conditions, Ω^+ remains a simply connected domain and leans toward to the part of boundaries with high curvature. For Robin boundary conditions $\frac{\partial u}{\partial n} + \beta u = 0$, there exists a threshold β^* at which a transition from a Dirichlet-like scenario to Neumann-like one occurs.

In [7], the model (1) with $\tau = 0$ and without the reaction term $f(x, u)$ was analyzed for longtime behavior. The authors proved that similar to the heat equation, the solution exponentially converges to zero for Dirichlet boundary conditions and exponentially converges to the mean value of the initial condition for Neumann conditions. In [31], proofs of the existence and uniqueness of positive solutions for the nonlocal dispersal equation were obtained by the monotone iteration method. A recent result on the optimal distribution of resources when the total resource was fixed was obtained in [1]. The authors proved that the optimal $m(x)$ is of bang-bang type for the model (1) with $\tau = 0$ and a uniform kernel on a bounded domain $\Omega \subset \mathbb{R}$.

The stability of the equilibrium solution $u = 0$ of (7) and (8) depends on the sign of the principal eigenvalue of the corresponding linear eigenvalue problem

$$\mathscr{L}\phi \equiv -d\left[\tau \Delta \phi + (1 - \tau)\,\mathscr{K}\phi\right] - m(x)\phi = \lambda\phi. \tag{9}$$

with the corresponding boundary conditions. It was shown in [24] that a variational formulation can be used for the principal eigenvalue

$$\lambda_p = \inf_{0 \neq v \in \mathscr{H}^1(\Omega)} \frac{\int_{\Omega = (0, p_1) \times (0, p_2) \times \cdots \times (0, p_N)} v\mathscr{L}v dx}{\int_{\Omega = (0, p_1) \times (0, p_2) \times \cdots \times (0, p_N)} v^2 dx} \tag{10}$$

for periodic boundary conditions and

$$\lambda_D = \inf_{0 \neq v \in \mathscr{H}_0^1(\Omega)} \frac{\int_{\Omega} v\mathscr{L}v dx}{\int_{\Omega} v^2 dx} \tag{11}$$

for Dirichlet boundary conditions.

When the principal eigenvalue is positive, $u = 0$ is a stable equilibrium and the species cannot invade from a low initial population. When the principal eigenvalue is negative, $u = 0$ is unstable and the species can invade from a low initial population. The interesting question is how the spatial heterogeneity of the resource which is represented by $m(x)$ can affect the principal eigenvalue.

Thus, we will study the optimization problem which looks for the minimal eigenvalue

$$\lambda_p^* = \min_{m(x)} \lambda_p \text{ and } \lambda_D^* = \min_{m(x)} \lambda_D, \tag{12}$$

when the resource function $m(x)$ is a bang-bang function subject to

$$m = -m_1 \text{ or } m_2 \text{ a.e. and } \int_\Omega m(x)dx = M. \tag{13}$$

The rest of this paper is organized as follows. Section 2 is devoted to the analytic study of the mixed model with Dirichlet boundary conditions. In Section 3, we present numerical approaches to solve time-dependent equations (7) and (8). We also study the corresponding linearized eigenvalue problem (9) and show how the eigenvalue varies with respect to diffusion coefficients d and τ. Furthermore, we use a rearrangement algorithm to find the optimal configuration of $m(x)$ in (12) which minimizes the principal eigenvalue so that the species will die out more slowly or survive more easily. Conclusions and future work are discussed in Section 5.

2 Analytical Results

In [24], the time evolution problem with periodic boundary conditions (8) was studied and the solution was shown to be nonnegativity preserving (i.e., if the initial data $u(x, 0)$ is nonnegative, the solution $u(x, t)$ remains nonnegative at all later times t). For any value of the parameter $\delta \neq 0$, the integral operator $\mathscr{K}$ is of Hilbert–Schmidt type (that implies it is continuous and compact). $\mathscr{K}$ is a self-adjoint operator $L^2(\Omega) \to L^2(\Omega)$ [24].

In the following, we will focus on the proofs for the properties of mixed dispersal model with Dirichlet boundary conditions.

Theorem 1 *There exists a constant D^*, such that if the diffusion coefficient $d > D^*$, no positive stationary states of (1) exist.*

Proof Let us first estimate the integration involving $\mathscr{K}$:

$$\int_\Omega u\,(\mathscr{K}u)\,dx = \int_\Omega \int_\Omega u(x)k(|x-y|)u(y)dydx - \int_\Omega u^2(x)dx$$

$$= \int_\Omega \int_\Omega u(x)k(|x-y|)u(y)dydx - \int_\Omega k(|x-y|)dy \int_\Omega u^2(x)dx$$

$$= \int_\Omega \int_\Omega u(x)\left[k(|x-y|)\,(u(y)-u(x))\right]dydx$$

$$= \frac{1}{2}\int_\Omega \int_\Omega u(x)\left[k(|x-y|)\,(u(y)-u(x))\right]dydx$$

$$+ \frac{1}{2} \int_{\Omega} \int_{\Omega} u(y) \left[k(|y - x|) (u(x) - u(y)) \right] dx dy$$

$$= -\frac{1}{2} \int_{\Omega} \int_{\Omega} \left[k(|x - y|) (u(y) - u(x))^2 \right] dy dx$$

$$\leq 0 . \tag{14}$$

Let $u(x)$ be a nonnegative stationary solution of (7). Then $u(x)$ satisfies

$$\tau \Delta u = -\frac{1}{d} u(m(x) - u) + (\tau - 1) \mathcal{K} u.$$

Multiplying the equation by u and integrating over the domain Ω, we obtain

$$\tau \int_{\Omega} u \Delta u dx = -\frac{1}{d} \int_{\Omega} u^2 (m(x) - u) dx + (\tau - 1) \int_{\Omega} u (\mathcal{K} u) dx.$$

Integrating by parts and using the Dirichlet boundary conditions, we obtain

$$\tau \int_{\Omega} |\nabla u|^2 dx = \frac{1}{d} \int_{\Omega} u^2 (m(x) - u) dx + (1 - \tau) \int_{\Omega} u (\mathcal{K} u) dx$$

$$\leq \frac{1}{d} \int_{\Omega} u^2 m(x) dx \leq \frac{m_2}{d} \int_{\Omega} u^2 dx.$$

Applying the Poincare inequality on the right-hand-side term, we obtain

$$\tau \int_{\Omega} |\nabla u|^2 dx \leq c_{\Omega} \frac{m_2}{d} \int_{\Omega} |\nabla u|^2 dx$$

where c_{Ω} is a constant that depends on Ω. In one dimension, $c_{\Omega} = \frac{|\Omega|^2}{\pi^2}$. Thus

$$(d\tau - c_{\Omega} m_2) \int_{\Omega} |\nabla u|^2 dx \leq 0.$$

If

$$d \geq \frac{m_2 c_{\Omega}}{\tau} =: D^*, \tag{15}$$

it implies that we must have $u \equiv 0$. $\qquad\qquad\qquad\qquad\qquad\qquad\qquad\qquad\square$

Theorem 2 *If $d > D^*$, then $\|u\|_1$ decays exponentially.*

Proof Let us define the energy function $E(t) = \frac{1}{2} \int_{\Omega} u^2(x, t) dx$. The rate of change of energy is

$$E'(t) = \int_\Omega u u_t dx$$

$$= d \left(\tau \int_\Omega u \Delta u dx + (1 - \tau) u \left(\mathcal{K} u \right) dx \right) + \int_\Omega m u^2 dx - \int_\Omega u^3 dx.$$

Dropping the last term due to the positivity preserving of the solution [24, 29], applying the kernel estimation (14) on the second term, and applying the Poincare inequality on the third term, we obtain

$$E'(t) \le [-d\tau + m_2 c_\Omega] \int_\Omega |\nabla u|^2 dx.$$

Denote $D := [-d\tau + m_2 c_\Omega]$. Note that $d > D^*$ implies that $D < 0$. Applying Poincare inequality again, we then have

$$E'(t) \le D \int_\Omega |\nabla u|^2 \le \frac{D}{c_\Omega} \int_\Omega u^2 dx.$$

By using Grönwall's inequality, now we have

$$E(t) \le E_0 \exp \left(\frac{2D}{c_\Omega} t \right).$$

As a result, we now have the decay rate estimation of 1-norm of $u(x, t)$:

$$\int_\Omega |u| dx \le |\Omega|^{1/2} \left(\int_\Omega u^2 dx \right)^{1/2} = 2^{1/2} |\Omega|^{1/2} E^{1/2}(t)$$

$$\le (2|\Omega| E_0)^{1/2} \exp \left(\frac{D}{c_\Omega} t \right). \qquad \Box$$

Theorem 3 *If $d < D^*$, $\|u\|_1$ is bounded from above.*

Proof Multiplying equation (7) by u and integrating over the domain Ω, we obtain

$$\int_\Omega u^3 dx = -d\tau \int_\Omega |\nabla u|^2 dx + d(1 - \tau) \int_\Omega u \left(\mathcal{K} u \right) dx + \int_\Omega m(x) u^2 dx$$

which implies that

$$\|u\|_3^3 = \int_\Omega u^3 dx \le \left(m_2 - \frac{d\tau}{c_\Omega} \right) \int_\Omega u^2 dx.$$

By using $\|u^2\|_{3/2} = \|u\|_3^2$, we have

$$\int_\Omega u^2 dx \le \left(\int_\Omega (u^2)^{3/2} dx \right)^{3/2} |\Omega|^{1/3} \le \|u\|_3^2 |\Omega|^{1/3}.$$

Thus,

$$\|u\|_3 \le \left(m_2 - \frac{d\tau}{c_\Omega}\right) |\Omega|^{1/3}.$$

We then have

$$\|u\|_1 = \int_\Omega u dx \le \left(\int_\Omega u^3 dx\right)^{1/3} |\Omega|^{2/3}$$

$$\le \left(m_2 - \frac{d\tau}{c_\Omega}\right)^{1/3} |\Omega|^{7/9}. \qquad \square$$

Note that the upper bound derived here is a decreasing function in d.

Theorem 4 *If $d > D^*$, then the principal eigenvalue is positive if exists.*

Proof Denote the normalized eigenfunction, i.e., $\int_\Omega \phi^2 dx = 1$, which corresponds to λ_D by ϕ, thus

$$\lambda_D = -\int_\Omega \phi \{d \, [\tau \Delta \phi + (1 - \tau) \mathcal{K} \phi] + m\phi\} \, dx$$

$$= \int_\Omega d \left[\tau |\nabla \phi|^2 - (1 - \tau)\phi \, (\mathcal{K} \phi)\right] - m\phi^2 dx$$

$$\ge d \left[\frac{\tau}{c_\Omega} - \frac{m_2}{d}\right] \int_\Omega \phi^2 dx = d \left[\frac{\tau}{c_\Omega} - \frac{m_2}{d}\right].$$

Thus if $d > D^*$ defined in (15), we have $\lambda_D > 0$. $\qquad \square$

Theorem 5 *Let $\lambda_D(d_1)$ and $\lambda_D(d_2)$ be the corresponding principal eigenvalues for $d = d_1$ and $d = d_2$. If $d_1 > d_2$, then $\lambda_D(d_1) > \lambda_D(d_2)$.*

Proof Denote the normalized eigenfunction which corresponds to $\lambda_D(d_1)$ by ϕ, thus

$$- \{d_1 \, [\tau \Delta \phi + (1 - \tau) \mathcal{K} \phi] + m\phi\} = \lambda_D(d_1)\phi. \qquad (16)$$

Since ϕ is not necessarily the eigenfunction which corresponds to $\lambda_D(d_2)$, we have

$$- \int_\Omega \{d_2 \, [\tau \Delta \phi + (1 - \tau) \mathcal{K} \phi] + m\phi\} \, \phi dx \ge \lambda_D(d_2). \qquad (17)$$

Multiplying (16) by ϕ, integrating by parts, and subtracting (17), we have

$$\lambda_D(d_1) - \lambda_D(d_2) \ge (d_1 - d_2) \left[\tau \int |\nabla \phi|^2 dx - (1 - \tau) \int \phi \, (\mathcal{K} \phi) \, dx\right] > 0.$$

$\square$

Theorem 6 *Let $\lambda_D(\tau_1)$ and $\lambda_D(\tau_2)$ be the corresponding principal eigenvalues for $\tau = \tau_1$ and $\tau = \tau_2$. For any given Ω with Poincare constant $c_\Omega < 1$, if $\tau_1 > \tau_2$, then $\lambda_D(\tau_1) > \lambda_D(\tau_2)$.*

Proof Denote the normalized eigenfunction which corresponds to $\lambda_D(\tau_1)$ by ϕ, thus

$$-\{d\,[\tau_1\Delta\phi + (1-\tau_1)\mathscr{K}\phi] + m\phi\} = \lambda_D(\tau_1)\phi. \tag{18}$$

Since ϕ is not necessary the eigenfunction which corresponds to $\lambda_D(\tau_2)$, we have

$$-\int_\Omega \{d\,[\tau_2\Delta\phi + (1-\tau_2)\mathscr{K}\phi] + m\phi\}\,\phi dx \geq \lambda_D(\tau_2). \tag{19}$$

Multiplying (18) by ϕ, integrating by parts, and subtracting (19), we have

$$\lambda_D(\tau_1) - \lambda_D(\tau_2) \geq d(\tau_1 - \tau_2)\left[\int |\nabla\phi|^2 dx + \int \phi\,(\mathscr{K}\phi)\,dx\right]$$

$$\geq d(\tau_1 - \tau_2)\left[\frac{1}{c_\Omega}\int \phi^2 dx - \int \phi^2 dx\right]$$

$$\geq d(\tau_1 - \tau_2)\left[\frac{1}{c_\Omega} - 1\right]\int \phi^2 dx = d(\tau_1 - \tau_2)\left[\frac{1}{c_\Omega} - 1\right].$$

If $c_\Omega < 1$, we have $\lambda_D(\tau_1) > \lambda_D(\tau_2)$. $\qquad\square$

Theorem 7 *Let $m(x)$ be any given function satisfying (5) and $\bar{m}(\phi)$ be the bang-bang function satisfying (13) for the normalized eigenfunction $\phi(x)$ of $m(x)$, i.e., $\bar{m}(\phi) = m_2\chi_{E_\alpha} - m_1\chi_{\Omega\backslash E_\alpha}$ where $E_\alpha(\phi) = \{x \in \Omega : \phi^2(x) > \alpha\}$ and α is chosen so that $|E_\alpha|m_2 - |\Omega\backslash E_\alpha|m_1 = M$ holds. (In case $\phi = 0$ on a set of positive measure, it is also possible to choose such a bang-bang function. See [4].) Then, the principal eigenvalue satisfies*

$$\lambda_D(m) \geq \lambda_D(\bar{m}).$$

Proof For any given eigenfunction ψ,

$$\int_\Omega (\bar{m}(\psi) - m)\,\psi^2 dx = \int_{E_\alpha} (m_2 - m)\,\psi^2 dx + \int_{\Omega\backslash E_\alpha} (-m_1 - m)\,\psi^2 dx$$

$$\geq \alpha\int_{E_\alpha} (m_2 - m)\,dx - \alpha\int_{\Omega\backslash E_\alpha} (m_1 + m)\,dx$$

$$= \alpha\int_{E_\alpha} (\bar{m} - m)\,dx = 0,$$

we obtain

$$\lambda_D(\bar{m}) = \inf_{0\neq v\in H_0^1,\,||v||_2=1} \int_\Omega -d\left[-\tau|\nabla v|^2 + (1-\tau)\,v(\mathscr{K}v)\right] - \bar{m}v^2 dx$$

$$\leq \int_{\Omega} -d\left[-\tau|\nabla\phi|^2 + (1-\tau)\,\phi(\mathcal{K}\phi)\right] - \bar{m}\phi^2 dx$$

$$\leq \int_{\Omega} -d\left[-\tau|\nabla\phi|^2 + (1-\tau)\,\phi(\mathcal{K}\phi)\right] - m(x)\phi^2 dx$$

$$= \lambda_D(m(x))$$

where ϕ is the normalized eigenfunction corresponds to $m(x)$. The proof for $\lambda_p(\bar{m}) \leq \lambda_p(m)$ follows the same arguments.   □

Here, we prove a theorem which is related to our rearrangement algorithm to find the optimal configuration of the resource function $m(x)$. Given a function $m(x)$ defined in Ω and satisfying (5), we say that $m_0(x)$ belongs to the class of rearrangements $\mathcal{R} = \mathcal{R}(m(x))$ if

$$\text{mes}\{x \in \Omega : m_0(x) \geq \beta\} = \text{mes}\{x \in \Omega : m(x) \geq \beta\}, \quad \forall \beta \geq 0.$$

Theorem 8 *Let $m_0(x)$ belongs to the class of rearrangements $\mathcal{R}(m(x))$. Denote the corresponding normalized eigenfunctions of resource functions $m(x)$ and $m_0(x)$ by $\phi(x)$ and $\phi_0(x)$, respectively. If*

$$\int_{\Omega} m_0\phi^2 dx \geq \int_{\Omega} m\phi^2 dx, \tag{20}$$

then $\lambda_D(m_0) \leq \lambda_D(m)$. Similarly, $\lambda_p(m_0) \leq \lambda_p(m)$.

Proof By the definition of the principal eigenvalue, we have

$$\lambda_D(m_0(x)) = \int_{\Omega} -d\left[-\tau|\nabla\phi_0|^2 + (1-\tau)\,\phi_0(\mathcal{K}\phi_0)\right] - m_0(x)\phi_0^2 dx$$

$$\leq \int_{\Omega} -d\left[-\tau|\nabla\phi|^2 + (1-\tau)\,\phi(\mathcal{K}\phi)\right] - m_0(x)\phi^2 dx$$

$$\leq \int_{\Omega} -d\left[-\tau|\nabla\phi|^2 + (1-\tau)\,\phi(\mathcal{K}\phi)\right] - m(x)\phi^2 dx$$

$$= \lambda_D(m(x)).$$

The proof for $\lambda_p(m_0) \leq \lambda_p(m)$ follows the same arguments.   □

This result allows one to find a new configuration $m_0(x)$ satisfying (13) with a smaller eigenvalue for any given $m(x)$ satisfying (13). It is well known that $\sup_{m(x)} \int_{\Omega} m(x)\phi^2 dx$ is obtained when $m(x)$ is arranged to be a monotone increasing function in ϕ^2 [19] which means that the optimal choice is $\bar{m}(\phi) = m_2\chi_{E_\alpha} - m_1\chi_{E_{\Omega\setminus\alpha}}$ where $E_\alpha(\phi) = \left\{x \in \Omega : \phi^2(x) > \alpha\right\}$ and α is chosen so that $|E_\alpha|m_2 - |\Omega\setminus E_\alpha|m_1 = M$ holds.

3   Numerical Implementation

In this section, we discuss the numerical approaches to the mixed dispersal model with both Dirichlet and periodic boundary conditions. We solve the linearized eigenvalue problem (9) and time evolution problems (7) and (8), as well as the optimization problem which determines the optimal arrangement of the resources. For simplicity, we use finite difference approaches on one-dimensional intervals and two-dimensional rectangular domains and use a finite element approach on general domains such as dumbbell shapes with Dirichlet boundary conditions.

3.1   Finite Difference Method

For one-dimensional finite difference approach, we use the method proposed in [24]. The model (1) in one-dimensional interval $I = [0, L]$ is described by

$$u_t = d \left[\tau u_{xx} + (1 - \tau) \left(\frac{1}{\delta} \int_I k^* \left(\frac{x - y}{\delta} \right) u(y) dy - u(x) \right) \right] + u(m(x) - u), \tag{21}$$

with k^* defined in (6). The corresponding eigenvalue problem is

$$-d \left[\tau u_{xx} + (1 - \tau) \left(\frac{1}{\delta} \int_I k^* \left(\frac{x - y}{\delta} \right) u(y) dy - u(x) \right) \right] - m(x) u = \lambda u. \tag{22}$$

We discretize the domain by using a uniform mesh: $x_i = ih$ for $i = 0, 1, 2, \ldots, N$, with the mesh size $h = L/N$. Denote by u_i the numerical approximation of $u(x_i)$. For Dirichlet boundary conditions, zero values are assigned to u_0 and u_N. We seek for the solution $\mathbf{U} = (u_1, u_2, \ldots, u_{N-1})^T$. For periodic boundary conditions, we have $u_0 = u_N$ and seek for the solution $\mathbf{U} = (u_1, \ldots, u_N)^T$. The local dispersal term u_{xx} is approximated by a second order central difference scheme and the nonlocal dispersal term involving integration is approximated by the trapezoidal method. The discretization of (21) leads to a system of ordinary differential equations. With given initial values of $\mathbf{U}$, we use forward Euler method to compute the solution at any later time under the stability restriction on the time stepsize. The discretization of (22) leads to a discrete eigenvalue problem and the principal eigenvalue can be easily computed using Arnoldi's method.

It is very straightforward to extend this method to solve (7) and (8) in two dimensions on a rectangular domain $[0, L_x] \times [0, L_y]$. We discretize the domain by using a mesh: $(x_i, y_j) = (ih_x, jh_y)$ for $i = 0, 1, 2, \ldots, N_x$ and $j = 0, 1, 2, \ldots, N_y$, with $h_x = L_x/N_x$ and $h_y = L_y/N_y$. Denote by $u_{i,j}$ the numerical approximation of $u(x_i, y_j)$. Boundary conditions are enforced in the similar way as the ones in one dimension. The local dispersal term Δu is approximated by the five-point difference scheme

$$\Delta u(x_i, y_j) \approx \frac{u_{i+1,j} - 2u_{i,j} + u_{i-1,j}}{h_x^2} + \frac{u_{i,j+1} - 2u_{i,j} + u_{i,j-1}}{h_y^2}$$

and the nonlocal dispersal term involving integration of the kernel term

$$\int_0^{L_x} \int_0^{L_y} k(|(x, y) - (\tilde{x}, \tilde{y})|) u(\tilde{x}, \tilde{y}) d\tilde{x} d\tilde{y}$$

is done with a composite trapezoidal rule in two dimensions

$$\sum \sum k_{i,j,\tilde{i},\tilde{j}} w_{\tilde{i},\tilde{j}} u_{\tilde{i},\tilde{j}}$$

where $k_{i,j,\tilde{i},\tilde{j}}$ is $k(|(x_i, y_j) - (\tilde{x}_i, \tilde{y}_j)|)$ and $w_{\tilde{i},\tilde{j}}$ are weights of composite trapezoidal rule. The discretization of (7) and (8) again leads to a system of ordinary differential equations while the discretization of (9) leads to a discrete eigenvalue problem.

3.2 Finite Element Method

On irregular domains, we use a finite element method to solve (9) with Dirichlet boundary conditions. We use bilinear elements on quadrilaterals for the eigenfunction, and the function $m(x)$ is represented by piecewise constants on the quadrilaterals. The discrete eigenvalue problem

$$PU = \lambda QU \tag{23}$$

where U is a solution vector with entries U_j ($j = 1, \ldots, n$), P is the matrix resulting from the differential operator, integral operator, and the resource function term, and Q is the mass matrix, is solved by using Arnoldi's algorithm. We use the deal.II finite element library [3] to do our computations.

Algorithm 1 A rearrangement algorithm to minimize the principal eigenvalue

---
Give an initial guess for $m(x)$ and compute the area of favorable region $|\Omega_+|$.
Repeat 1-3 until $m(x)$ does not change any more
   1. Solve the eigenvalue problem (9) with Dirichlet or periodic boundary conditions by the finite difference method described in Section 3.1 or the finite element method described in Section 3.2.
   2. Sort the value of ϕ^2 at the discrete points in the descending order and compute the threshold α such that $|\{x|\phi(x)^2 > \alpha\}| = |\Omega_+|$.
   3. If $\phi(x)^2 >= \alpha$, assign $m(x) = m_2$. Otherwise, assign $m(x) = -m_1$.

3.3 Optimization Approach Based on a Rearrangement Algorithm

For finding the minimal principal eigenvalue in (12) subjected to (13), we adopt the rearrangement algorithm proposed in [26]. Given an initial configuration of $m(x)$ which is bounded from below by $-m_1$ and bounded from above by m_2, one can calculate the area for the favorable region Ω_+. In the finite difference calculation, this is done by counting the number of the mesh points which have the value $m = m_2$ and multiplying by h in one dimension and by $h_x h_y$ in two dimensions. In the finite element approach, this is done by calculating the total area of the elements which have the value $m = m_2$. Next, using the rearrangement approach, we iterate $m(x)$ until the optimal configuration is reached. In each iteration, m_2 is assigned to the location where square values of magnitude of the eigenfunction is above the critical threshold α which is chosen when $|\{x | \phi(x)^2 > \alpha\}| = |\Omega_+|$ is satisfied. The algorithm is summarized in Algorithm 1. The stopping criterion is when two successive $m(x)$ are identically the same. If the algorithm stops at $n-$th iteration, it means that the optimal configuration is achieved at $(n-1)-$th iteration.

4 Numerical Results

In this section, we study how the principal eigenvalue changes for different values of coefficients d and τ, for both Dirichlet and periodic boundary conditions. Results from computations on square and rectangular domains in one and two dimensions are presented. We also present results of simulations on a general-shaped domain in two dimensions.

4.1 One-Dimensional Results

We first study how the principal eigenvalue of Equation (22) in the interval $[0, 1]$ varies with respect to different values of coefficients, d and τ. In our experiments, we use mesh size $N = 400$ for periodic boundary conditions and $N = 1600$ for Dirichlet boundary conditions to guarantee at least two significant digits of accuracy for the principal eigenvalue. Given $m(x) = \chi_{[0.4,0.6]} - \chi_{[0,0.4) \cup (0.6,1]}$, we compute the principal eigenvalue for d ranging from 0 to 0.1 with $\tau = 0, 0.5$, and 1. In Figure 1(a) and 1(b), we see that the principal eigenvalue becomes negative when the total diffusion coefficient d is relatively small in both periodic and Dirichlet cases. Furthermore, the principal eigenvalue becomes smaller as τ decreases. In Figure 2(a) and 2(b), the change of λ_p and λ_D are shown with respect to τ ranging from 0 to 1 when $d = 0.01, 0.05, 0.1$, and 1. In both periodic and Dirichlet cases, we see that λ_p and λ_D become negative when τ is relatively small, which means the

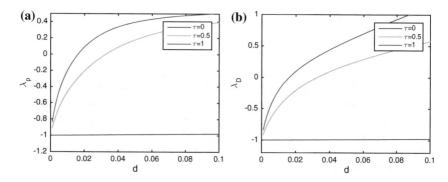

Fig. 1 (a) λ_p versus d (b) λ_D versus d for $\tau = 0, 0.5$, and 1 with $\delta = 0.15$.

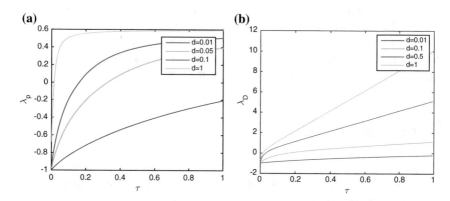

Fig. 2 (a) λ_p versus τ (b) λ_D versus τ for $d = 0.01, 0.05, 0.1$ and 1 with $\delta = 0.15$.

species adopted more nonlocal diffusion. Furthermore, λ_p and λ_D are both monotone increasing functions in d with the current choice of parameters.

For time evolution problems (7) and (8), we study the long-term behavior of the solution. When the diffusion coefficient d is big enough, for example see Figure 3 with $d = 2$, the whole population dies out as time goes to infinity. In Figure 3(a) and 3(b), we plot the logarithm of the L_1 norm of u for both periodic and Dirichlet boundary conditions, upon which we also perform linear fitting, shown in Figure 3(c) and 3(d). We can see that $||u||_{L_1}$ decays exponentially for both boundary conditions, which is consistent with our analytical result in Theorem 2. On the other hand, for small total diffusion d, the solution will finally reach a positive steady state for both periodic and Dirichlet boundary conditions, as illustrated in Figure 4 when $d = 0.02$, where we plot the final configuration of u after time long enough to show the positive steady state.

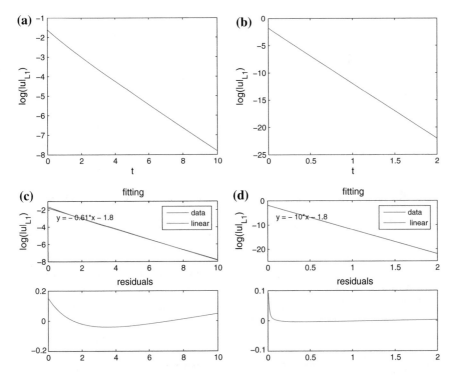

Fig. 3 For large d, the population vanishes with exponential decay. In the above experiment, $d = 2$, $\tau = 0.5$, and timestep $= 0.4h^2$. (a) and (b) show the logarithm of the L_1 norm of u versus time. (c) and (d) are the linear fittings and the residues. The left column is for periodic boundary conditions and the right column is for Dirichlet boundary conditions.

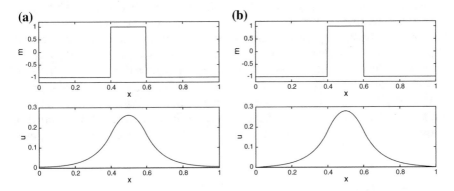

Fig. 4 For small d, the population reaches a positive steady state eventually. Here, $d = 0.02$, $\tau = 0.5$, $\delta = 0.15$, $N = 400$, and timestep $= 0.4h^2$. (a) and (b) show the final configurations of u. The left column is for periodic boundary conditions and the right column is for Dirichlet boundary conditions.

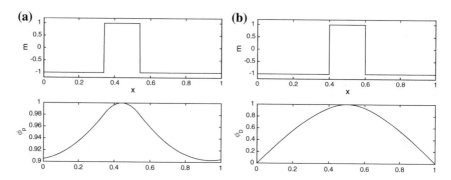

Fig. 5 The optimal configuration of $m(x)$ and the corresponding eigenfunction computed by the rearrangement approach with $d = 0.8$, $\tau = 0.5$, $\delta = 0.15$ for (a) periodic boundary conditions ($\lambda_p = 0.57$) and (b) Dirichlet boundary conditions ($\lambda_D = 4.16$).

Finally, we determine the optimal arrangement of $m(x)$ in an interval by using the rearrangement approach. It turns out that for both periodic and Dirichlet boundary conditions, the optimal favorable region is connected, as shown in Figure 5 with $d = 0.8$. Here, our initial guess of $m(x)$ is

$$m(x) = \begin{cases} m_2 \text{ for } |x - 0.4| < 0.05, |x - 0.8| < 0.05 \\ m_1 \text{ for } \qquad\qquad\text{otherwise} \end{cases}$$

which has two positive intervals with a total area to be 0.2. For Dirichlet boundary conditions in Figure 5(b), the optimal favorable region is found to be in the center of the domain with area 0.2, after 5 iterations. However, for periodic boundary conditions, the result is slightly different. For this particular initial m, after 3 iterations, the system reaches the optimization solution and the favorable region is shown in Figure 5(a), where the positive favorable region is still connected and with area 0.2, but in a different position. In fact, for periodic boundary conditions, the positive favorable region does not necessarily stay fixed on one part of the domain in the one-dimensional case. The connected positive favorable region is transferable to any part of the domain, as long as it has the same total area. Thus, the numerical optimization solution depends on the initial guess of m. From the results of our numerical experiments, we see that the rearrangement algorithm converges in a few iterations.

4.2 Two-Dimensional Results

In two dimensions, we perform similar numerical tests as in one dimension with $m_1 = m_2 = 1$. On a square domain $[0, 1] \times [0, 1]$, we set the mesh size $Nx = Ny = 50$ for both periodic and Dirichlet boundary conditions. The initial guess of favorable region of $m(x)$ is a circle centered at $(0.5, 0.5)$ with radius 0.3. Using exactly the

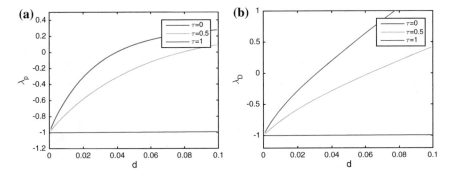

Fig. 6 (a) λ_p versus d (b) λ_D versus d with $\delta = 0.15$ for $\tau = 0, 0.5$, and 1 in the two-dimensional case.

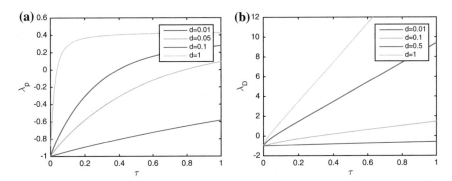

Fig. 7 (a) λ_p versus τ (b) λ_D versus τ with $\delta = 0.15$ for $d = 0.01, 0.05, 0.1$, and 1 in the two-dimensional case.

same parameters d and τ as in the one-dimensional case, we plot in Figure 6(a) and (b) how the principal eigenvalue changes with respect to d for periodic and Dirichlet boundary conditions. In Figure 7, we show the curves of λ_p and λ_D changing with respect to τ for different d. The results in two dimensions are very similar to those in one dimension. Both λ_p and λ_D get smaller as d or τ decreases with the current choice of parameters.

In order to find the optimal favorable region in two dimensions, we solve the optimization problem using a rearrangement approach. First, we study the problem on a square domain. We set the initial guess of $m(x)$ (Figure 8) to be a cross centered at $(0.3, 0.3)$ with the definition below.

$$m(x, y) = \begin{cases} m_2 & \text{for } |x - 0.3| < 0.1 \text{ and } |y - 0.3| < 0.2, \\ m_2 & \text{for } |x - 0.3| < 0.2 \text{ and } |y - 0.3| < 0.1, \\ m_1 & \text{for } \quad \text{otherwise.} \end{cases} \tag{24}$$

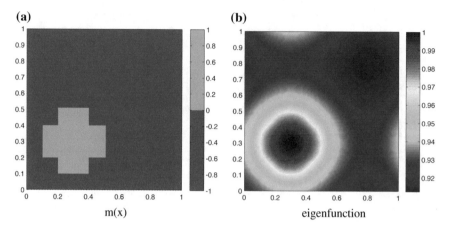

Fig. 8 (a) Initial configuration of $m(x)$ with a cross-shaped favorable region for the optimization problem and (b) its corresponding eigenfunction with periodic boundary conditions in two dimensions.

The total area of the favorable region of the initial $m(x)$ is 0.12. For periodic boundary conditions, the optimized solution is reached after 3 iterations. In Figure 9, we show the configuration of $m(x)$ and its corresponding eigenfunction at each iteration. As we can see, the optimal shape of the favorable region of m turns into a circular-like shape during evolution.

For Dirichlet boundary conditions, the optimal result is also obtained after 3 iterations, as shown in Figure 10. The favorable region again turns into a circular-like shape but goes to the center of the domain, which is different from the periodic boundary case. As the number of iterations increases, λ_D decreases before settling at 10.07 in the end. In Table 1, we show the values of λ_D for each iteration for both periodic and Dirichlet boundary conditions.

In Figure 11, we show the optimal results at 4-th iterations with negative eigenvalues when the choices of parameters are $d = 0.01$, $\tau = 0.5$, $\delta = 0.15$ for both periodic and boundary conditions. The initial guess for periodic boundary condition is (24) and the initial guess for Dirichlet boundary condition is (24) with the center of the cross shifted from $(0.3, 0.3)$ to $(0.5, 0.5)$. This choice of the initial condition gives the optimal configuration with fewer iterations. We observe that the results are similar to the optimal configuration with positive eigenvalues.

Next, we explore the results on rectangular domains with the same width, 1, but with different heights, b. We set the initial guess of favorable region of m to be a square with width 0.2 centered at $(0.3, b/2)$. We calculate the optimal configurations of $m(x)$ for $b = 0.35$, 0.4, 0.5, and 0.8, for both periodic and Dirichlet boundary conditions, and present the final optimal m and the corresponding eigenfunctions in Figure 12 for periodic boundary conditions and in Figure 13 for Dirichlet boundary conditions. As we can see, for a periodic boundary, the optimal favorable region tends to a vertical band and touches the horizontal boundary for small b, and becomes a circular-like shape when b turns bigger, which does not necessarily stay in the middle

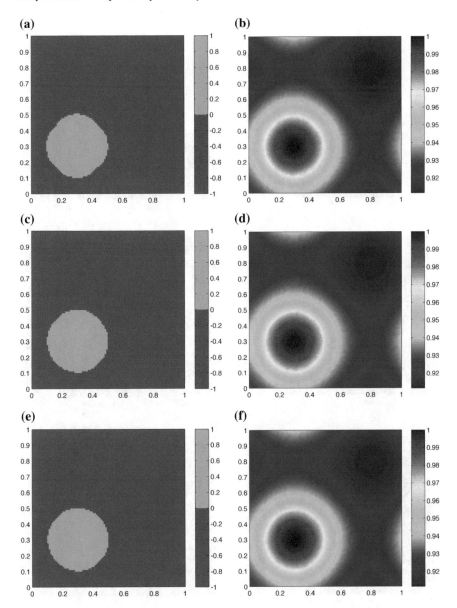

Fig. 9 The configuration of $m(x)$ and its corresponding eigenfunction for the first three numerical iterations. The optimal result $\lambda_p^* = 0.74$ is achieved at the third iteration. The choices of parameters are $d = 1$, $\tau = 0.5$, $\delta = 0.15$, and $N_x = N_y = 100$. (a) Iteration 1. (b) Iteration 1. (c) Iteration 2. (d) Iteration 2. (e) Iteration 3. (f) Iteration 3.

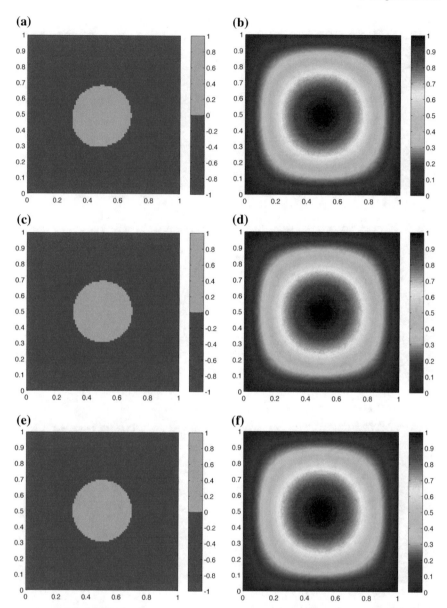

Fig. 10 The configuration of $m(x)$ and its corresponding eigenfunction for the first three numerical iterations. The optimal result $\lambda_D^* = 10.07$ is achieved at the third iteration. The choices of parameters are $d = 1$, $\tau = 0.5$, $\delta = 0.15$, and $N_x = N_y = 100$. (a) Iteration 1. (b) Iteration 1. (c) Iteration 2. (d) Iteration 2. (e) Iteration 3. (f) Iteration 3.

Table 1 Values of λ_p and λ_D at each iteration during optimization procedure for $d = 1$, $\tau = 0.5$, $\delta = 0.15$, $N_x = N_y = 100$.

Iteration	Periodic boundary	Dirichlet boundary
1	0.7398	10.499
2	0.7394	10.073
3	0.7394	10.070
4	0.7394	10.070

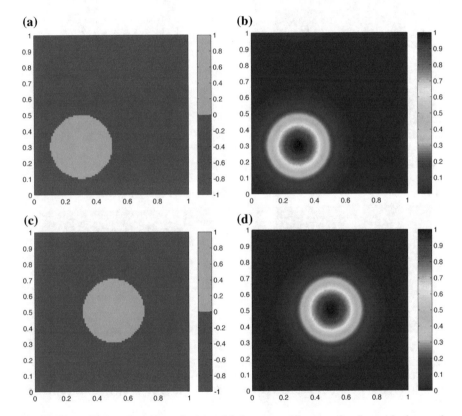

Fig. 11 The optimal configurations of $m(x)$ and their corresponding eigenfunctions at 4−th numerical iteration for both periodic (top row) and Dirichlet boundary conditions (bottom row). The optimal eigenvalues are $\lambda_p^* = -0.54$ and $\lambda_D^* = -0.54$. The choices of parameters are $d = 0.01$, $\tau = 0.5$, $\delta = 0.15$, and $N_x = N_y = 100$. (a) Iteration 4. (b) Iteration 4. (c) Iteration 4. (d) Iteration 4.

of the domain. However, for a Dirichlet boundary, for all choices of b examined, the positive resource tends to a circular-like shape and moves to the center of the domain.

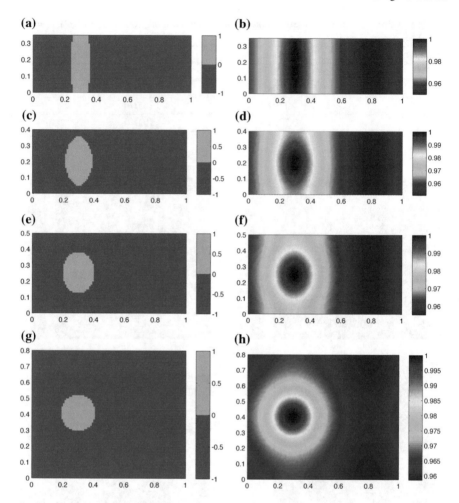

Fig. 12 The optimal favorable region and corresponding eigenfunction for $b = 0.35, 0.4, 0.5, 0.8$ for periodic boundary conditions. In these experiments, $d = 1$, $\tau = 0.5$, $\delta = 0.15$, $N_x = 100$, and $N_y = 100b$. (a) b = 0.35. (b) b = 0.35 (c) b = 0.4. (d) b = 0.4. (e) b = 0.5. (f) b = 0.5. (g) b = 0.8. (h) b = 0.8.

4.3 Dumbbell-Shaped Domain

Here, we present numerical results via a finite element method on a dumbbell-shaped domain with Dirichlet boundary conditions. The choices of parameters are $d = 1$, $\tau = 0.5$, and $\delta = 0.15$. We are looking for an optimal configuration of the resource function $m(x)$ that minimizes the principal eigenvalue of (9). In the first experiment, the domain Ω is dumbbell-shaped and the favorable region (corresponding to Ω_+) consists of two disks with the same radius located on both ends of the dumbbell.

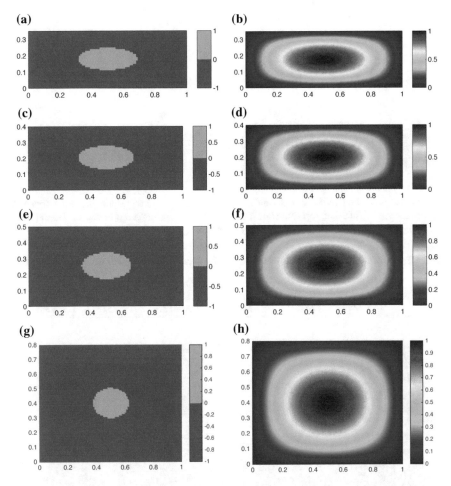

Fig. 13 The optimal favorable region and corresponding eigenfunction for $b = 0.35, 0.4, 0.5, 0.8$ for Dirichlet boundary conditions with $d = 1$, $\tau = 0.5$, $\delta = 0.15$, $N_x = 100$, and $N_y = 100b$. (a) b = 0.35. (b) b = 0.35 (c) b = 0.4. (d) b = 0.4. (e) b = 0.5. (f) b = 0.5. (g) b = 0.8. (h) b = 0.8.

The initial and optimal (final) configurations of $m(x)$ with the eigenfunctions corresponding to the principal eigenvalue are shown in Figure 14. The number of cells (quadrilaterals) in the domain is 25941. Table 2 shows the principal eigenvalue λ_D versus the iteration number.

For the next experiment, the two disks for the initial favorable regions have different radii. Results of the computation in this case are shown in Figure 15. Table 3 shows the principal eigenvalue λ_D after each iteration. As shown in the figures above, when the channel connecting the dumbbells is thinner, the optimal configuration of $m(x)$ is a single favorable region in either side. For cases where the two favorable regions on either side have different areas, the final configuration of $m(x)$ is concentrated on the side that had larger area initially. Next, we take a dumbbell-shaped domain with

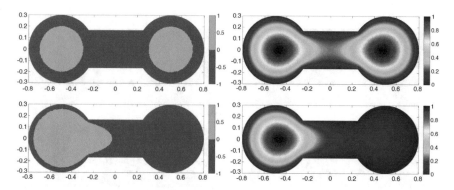

Fig. 14 Resource function $m(x)$ (left) and the eigenfunction $\phi_D(x)$ corresponding to the principal eigenvalue λ_D (right). The top row corresponds to the initial configuration and the bottom row corresponds to the optimal configuration after 9 iterations.

Table 2 The principal eigenvalue λ_D vs. iterations for a dumbbell-shaped domain (see Figure 14).

Iteration	0	2	4	6	8	10
λ_D	28.5249	28.4812	28.4797	28.4527	28.3513	28.3266

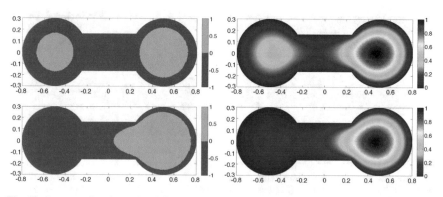

Fig. 15 Resource function $m(x)$ (left) and the eigenfunction $\phi_D(x)$ corresponding to the principal eigenvalue λ_D (right). The top row corresponds to the initial configuration and the bottom row corresponds to the optimal configuration after 3 iterations.

a thicker channel. Initial favorable regions are again two disks located at both ends. Results of the case where the two disks have the same radius are shown in Figure 16. Table 4 shows the principal eigenvalue λ_D after each iteration. The number of cells (quadrilaterals) in the domain is 21211. As we observe from the figure, the optimal favorable region is one large area in the middle of the thick channel away from the boundary. For the case when two disks that have different radii, we get a very similar configuration of $m(x)$.

Table 3 The principal eigenvalue λ_D vs. iterations for a dumbbell-shaped domain (see Figure 15).

Iteration	0	1	2	3	4
λ_D	28.5123	28.3965	28.3416	28.3334	28.3334

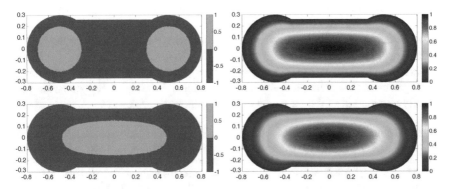

Fig. 16 Resource function $m(x)$ (left) and the eigenfunction $\phi_D(x)$ corresponding to the principal eigenvalue λ_D (right). The top row corresponds to the initial configuration and the bottom row corresponds to the optimal configuration after 4 iterations.

Table 4 The principal eigenvalue λ_D vs. iterations for a dumbbell-shaped domain with thick channel (see Figure 16).

Iteration	0	1	2	3	4	5
λ_D	20.0444	19.2431	19.2346	19.2346	19.2345	19.2345

5 Discussion

In this paper, we studied a mixed dispersal model with periodic and Dirichlet boundary conditions on different shapes of one-dimensional and two-dimensional domains both analytically and numerically. In terms of parameter values, we investigated two possible scenarios of longtime dynamics: the population of species dying off completely as time goes to infinity or converging to a nontrivial stationary distribution.

To analyze the convergence rate toward trivial and nontrivial stationary solutions, we estimated the principal eigenvalue of the corresponding linearized problem and solved the principal eigenvalue minimization problem in terms of the distribution of favorable and unfavorable regions. Our numerical simulations indicate that the optimal favorable region tends to be a simply connected domain. Numerous results are shown to demonstrate various scenarios of optimal favorable regions for periodic and Dirichlet boundary conditions.

We would expect similar results for more general Robin-type boundary conditions. It would be interesting to analyze how the longtime dynamics can be affected by the application of spectral-parameter- dependent boundary conditions (see [8] for example) or by replacing the linear diffusion term with nonlinear diffusion, taking into account that the diffusion coefficient depends on population density. For example, the diffusion coefficient can be taken proportional to some positive or negative power of density, hence embracing the cases where the diffusion coefficient grows or decreases with the density which will lead eventually to interesting pattern formations [9].

As a future work, we would also like to consider competition models when multiple species use different local and nonlocal dispersal strategies. Some results were already obtained in [23] on a model of two competing species.

Acknowledgments The authors would like to thank Institute for Mathematics and its Applications at University of Minnesota for hosting a special workshop on "WhAM! A Research Collaboration Workshop for Women in Applied Mathematics: Numerical Partial Differential Equations and Scientific Computing." This paper summarized the project results of Team3 on principal eigenvalue for a population dynamics model problem with both local and nonlocal dispersal.

References

1. Bai, X., Li, F.: Optimization of species survival for logistic models with non-local dispersal. Nonlinear Analysis: Real World Applications **21**, 53–62 (2014)
2. Bai, X., Li, F.: Global dynamics of a competition model with nonlocal dispersal ii: The full system. Journal of Differential Equations **258**(8), 2655–2685 (2015)
3. Bangerth, W., Heister, T., Heltai, L., Kanschat, G., Kronbichler, M., Maier, M., Turcksin, B., Young, T.D.: The deal.II library, version 8.2. Archive of Numerical Software **3** (2015)
4. Cantrell, R.S., Cosner, C.: Diffusive logistic equations with indefinite weights: population models in disrupted environments. Proceedings of the Royal Society of Edinburgh: Section A Mathematics **112**(3-4), 293–318 (1989)
5. Cantrell, R.S., Cosner, C.: Diffusive logistic equations with indefinite weights: population models in disrupted environments ii. SIAM Journal on Mathematical Analysis **22**(4), 1043–1064 (1991)
6. Cantrell, R.S., Cosner, C.: Spatial ecology via reaction-diffusion equations. John Wiley & Sons (2004)
7. Chasseigne, E., Chaves, M., Rossi, J.D.: Asymptotic behavior for nonlocal diffusion equations. Journal de mathématiques pures et appliquées **86**(3), 271–291 (2006)
8. Chugunova, M.: Inverse spectral problem for the sturm-liouville operator with eigenvalue parameter dependent boundary conditions. In: Operator theory, system theory and related topics, pp. 187–194. Springer (2001)
9. Colombo, E.H., Anteneodo, C.: Nonlinear diffusion effects on biological population spatial patterns. Physical Review E **86**(3), 036,215 (2012)
10. Cortazar, C., Coville, J., Elgueta, M., Martinez, S.: A nonlocal inhomogeneous dispersal process. Journal of Differential Equations **241**(2), 332–358 (2007)
11. Cortázar, C., Elgueta, M., García-Melián, J., Martínez, S.: Existence and asymptotic behavior of solutions to some inhomogeneous nonlocal diffusion problems. SIAM Journal on Mathematical Analysis **41**(5), 2136–2164 (2009)
12. Cosner, C., Dávila, J., Martínez, S.: Evolutionary stability of ideal free nonlocal dispersal. Journal of Biological Dynamics **6**(2), 395–405 (2012)

13. Coville, J.: On a simple criterion for the existence of a principal eigenfunction of some nonlocal operators. Journal of Differential Equations **249**(11), 2921–2953 (2010)
14. Coville, J., Dávila, J., Martínez, S.: Existence and uniqueness of solutions to a nonlocal equation with monostable nonlinearity. SIAM Journal on Mathematical Analysis **39**(5), 1693–1709 (2008)
15. Coville, J., Dávila, J., Martínez, S.: Nonlocal anisotropic dispersal with monostable nonlinearity. Journal of Differential Equations **244**(12), 3080–3118 (2008)
16. Coville, J., Davila, J., Martinez, S.: Pulsating fronts for nonlocal dispersion and kpp nonlinearity. Annales de l'Institut Henri Poincare (C) Non Linear Analysis **30**(2), 179–223 (2013)
17. Coville, J., Dupaigne, L.: On a non-local equation arising in population dynamics. Proceedings of the Royal Society of Edinburgh: Section A Mathematics **137**(04), 727–755 (2007)
18. Deng, K.: On a nonlocal reaction-diffusion population model. Discrete and Continuous Dynamical Systems Series B **9**(1), 65 (2008)
19. Henrot, A.: Extremum problems for eigenvalues of elliptic operators. Springer Science & Business Media (2006)
20. Hintermüller, M., Kao, C.Y., Laurain, A.: Principal eigenvalue minimization for an elliptic problem with indefinite weight and robin boundary conditions. Applied Mathematics and Optimization **65**(1), 111–146 (2012)
21. Jin, Y., Lewis, M.A.: Seasonal influences on population spread and persistence in streams: critical domain size. SIAM Journal on Applied Mathematics **71**(4), 1241–1262 (2011)
22. Jin, Y., Lewis, M.A.: Seasonal influences on population spread and persistence in streams: spreading speeds. Journal of Mathematical Biology **65**(3), 403–439 (2012)
23. Kao, C.Y., Lou, Y., Shen, W.: Random dispersal vs. nonlocal dispersal. Discrete and Continuous Dynamical Systems **26**(2), 551–596 (2010)
24. Kao, C.Y., Lou, Y., Shen, W.: Evolution of mixed dispersal in periodic environments. Discrete and Continuous Dynamical Systems B **17**(6), 2047–2072 (2012)
25. Kao, C.Y., Lou, Y., Yanagida, E.: Principal eigenvalue for an elliptic problem with indefinite weight on cylindrical domains. Mathematical Biosciences and Engineering **5**(2), 315–335 (2008)
26. Kao, C.Y., Su, S.: Efficient rearrangement algorithms for shape optimization on elliptic eigenvalue problems. Journal of Scientific Computing **54**(2-3), 492–512 (2013)
27. Li, F., Lou, Y., Wang, Y.: Global dynamics of a competition model with non-local dispersal i: The shadow system. Journal of Mathematical Analysis and Applications **412**(1), 485–497 (2014)
28. Lutscher, F.: Nonlocal dispersal and averaging in heterogeneous landscapes. Applicable Analysis **89**(7), 1091–1108 (2010)
29. Shen, W., Xie, X.: On principal spectrum points/principal eigenvalues of nonlocal dispersal operators and applications. arXiv preprint arXiv:1309.4753 (2013)
30. Skellam, J.: Random dispersal in theoretical populations. Biometrika pp. 196–218 (1951)
31. Sun, J.W.: Existence and uniqueness of positive solutions for a nonlocal dispersal population model. Electronic Journal of Differential Equations **2014**(143), 1–9 (2014)
32. Volpert, V., Vougalter, V.: Existence of stationary pulses for nonlocal reaction-diffusion equations. Documenta Mathematica **19**, 1141–1153 (2014)

Optimization-Based Decoupling Algorithms for a Fluid-Poroelastic System

Aycil Cesmelioglu, Hyesuk Lee, Annalisa Quaini, Kening Wang and Son-Young Yi

Abstract In this paper, computational algorithms for the Stokes-Biot coupled system are proposed to study the interaction of a free fluid with a poroelastic material. The decoupling strategy we employ is to cast the coupled fluid-poroelastic system as a constrained optimization problem with a Neumann type control that enforces continuity of the normal components of the stress on the interface. The optimization objective is to minimize any violation of the other interface conditions. Two numerical algorithms based on a residual updating technique are presented. One solves a least squares problem and the other solves a linear problem when the fluid velocity in the poroelastic structure is smooth enough. Both algorithms yield the minimizer of the constrained optimization problem. Some numerical results are provided to validate the accuracy and efficiency of the proposed algorithms.

A. Cesmelioglu
Department of Mathematics and Statistics, Oakland University,
Rochester, MI 48083, USA
e-mail: cesmelio@oakland.edu

H. Lee (✉)
Department of Mathematical Sciences, Clemson University,
Clemson, SC 29634, USA
e-mail: hklee@clemson.edu

A. Quaini
Department of Mathematics, University of Houston,
Houston, TX 77204, USA
e-mail: quaini@math.uh.edu

K. Wang
Department of Mathematics and Statistics,
University of North Florida, Jacksonville, FL 32224, USA
e-mail: kening.wang@unf.edu

S.-Y. Yi
Department of Mathematical Sciences,
The University of Texas at El Paso, El Paso, TX 79968, USA
e-mail: syi@utep.edu

© Springer Science+Business Media New York 2016
S.C. Brenner (ed.), *Topics in Numerical Partial Differential Equations
and Scientific Computing*, The IMA Volumes in Mathematics
and its Applications 160, DOI 10.1007/978-1-4939-6399-7_4

1 Introduction

We develop a robust and efficient numerical method to simulate the interaction of a free fluid with a deformable porous medium. Modeling fluid-poroelastic structure interaction is of great importance in a wide range of industrial and environmental applications, including groundwater flow, oil and gas production, blood-vessel interactions, breakwater design, and many more. See, e.g., [1–9] and references therein.

To model the free fluid, we consider the Stokes equations for a single phase, incompressible viscous fluid. A well accepted model for the fluid flow in a deformable porous medium is the Biot system [10–13]. The stress and flow couplings on the interface between the Biot flow through the deforming porous medium and the Stokes flow in the open channel must be prescribed by physically-consistent interface conditions. Refer to [14] for the formulation of the interface conditions and analytical study of the model.

The numerical discretization of the Stokes-Biot system poses great computational challenges due to the nature of its complexity. A fully-coupled scheme, which solves the Stokes and Biot subproblems simultaneously, results in a large linear system, which in turn requires a large amount of memory space and a special solver. The objective of this work is to develop efficient decoupling schemes that allow us to independently solve each subproblem using existing Stokes and Biot solvers possibly with slight modification, while ensuring convergence to an accurate solution. Recently, Bukač et al. [9] proposed and analyzed a loosely coupled scheme for the Stokes-Biot system, for which interface conditions are imposed for local problems by time lagging. In this work, we consider a different approach for decoupling, where the solution algorithm considered is based on optimization. We present a Neumann type control that enforces continuity of the normal components of the stress on the interface while minimizing any violation of the remaining interface conditions. Two numerical algorithms based on a residual updating technique are presented. One redefines the constrained optimization problem as a least squares problem whose solution yields the minimizer of the original constrained optimization problem. The other algorithm seeks the minimizer by solving a linear problem, assuming the fluid velocity in the poroelastic structure is smooth enough. Some numerical results are provided to validate the accuracy and efficiency of the proposed methods.

This paper is organized as follows. In Section 2 we present the governing equations of the Stokes-Biot problem, complemented by initial, boundary, and interface conditions. Time discretized weak formulation and its appropriate functional spaces are introduced in Section 3. Section 4 is devoted to the development of optimization-based decoupling schemes. Finally, in Section 5, we present some results of our numerical experiments.

2 Model Equations

Suppose that the domain under consideration is made up of two regions $\Omega^f(t) \in \mathbb{R}^2$ and $\Omega^p(t) \in \mathbb{R}^2$, t being time, separated by a common moving interface $\Gamma_{I(t)} = \partial \Omega^f(t) \cap \partial \Omega^p(t)$. See Figure 1. The first region $\Omega^f(t)$ is occupied by the free fluid and has boundary Γ^f such that $\Gamma^f = \Gamma^f_{in} \cup \Gamma^f_{out} \cup \Gamma^f_t \cup \Gamma_{I(t)}$, where Γ^f_{in} and Γ^f_{out} represent the inlet and outlet boundary, respectively. The second region $\Omega^p(t)$ is occupied by a saturated poroelastic structure with the boundary Γ^p such that $\Gamma^p = \Gamma^p_s \cup \Gamma^p_b \cup \Gamma_{I(t)}$, where $\Gamma^p_s \cup \Gamma^p_b$ represents the outer structure boundary.

For the sake of simplicity, we will consider the problem under the assumption of fixed domains $\Omega^f(t)$ and $\Omega^p(t)$. That is,

$$\Omega^f(t) = \Omega^f, \quad \Omega^p(t) = \Omega^p, \quad \Gamma_{I(t)} = \Gamma_I, \quad \forall t \in [0, T].$$

This assumption allows for a simple presentation of the algorithms to be proposed in the subsequent sections. In order to incorporate the full effect of the moving interface, we can employ the Arbitrary Lagrangian-Eulerian (ALE) formulation. Refer to [15] for a similar decoupling algorithm for fluid-structure interaction based on the ALE formulation.

Consider the fluid equations:

$$\rho_f \frac{\partial \mathbf{u}_f}{\partial t} - 2\nu_f \nabla \cdot D(\mathbf{u}_f) + \nabla p_f = \mathbf{f}_f \quad \text{in } \Omega^f, \tag{2.1a}$$

$$\nabla \cdot \mathbf{u}_f = 0 \quad \text{in } \Omega^f, \tag{2.1b}$$

where $\mathbf{u}_f$ denotes the velocity vector of the fluid, p_f the pressure of the fluid, ρ_f the density of the fluid, ν_f the fluid viscosity, and $\mathbf{f}_f$ the body force acting on the fluid. Here, $D(\mathbf{u}_f)$ is the strain rate tensor:

$$D(\mathbf{u}_f) = \frac{1}{2} \left(\nabla \mathbf{u}_f + (\nabla \mathbf{u}_f)^T \right).$$

Fig. 1 Fluid-poroelastic domain.

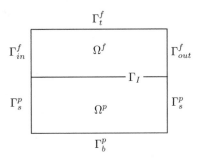

The Cauchy stress tensor is given by:

$$\boldsymbol{\sigma}_f = 2v_f D(\mathbf{u}_f) - p_f \mathbf{I}.$$

Equation (2.1a) represents the conservation of linear momentum, while equation (2.1b) represents the conservation of mass.

The poroelastic system is represented by the Biot model:

$$\rho_s \frac{\partial^2 \boldsymbol{\eta}}{\partial t^2} - 2v_s \nabla \cdot D(\boldsymbol{\eta}) - \lambda \nabla (\nabla \cdot \boldsymbol{\eta}) + \alpha \nabla p_p = \mathbf{f}_s \quad \text{in } \Omega^p , \tag{2.2a}$$

$$\kappa^{-1} \mathbf{u}_p + \nabla p_p = 0 \quad \text{in } \Omega^p , \tag{2.2b}$$

$$\frac{\partial}{\partial t} (s_0 p_p + \alpha \nabla \cdot \boldsymbol{\eta}) + \nabla \cdot \mathbf{u}_p = f_p \quad \text{in } \Omega^p , \tag{2.2c}$$

where $\boldsymbol{\eta}$ is the displacement of the structure, p_p is the pore pressure of the fluid, and $\mathbf{u}_p$ is the fluid velocity. Here, f_p is the source/sink term and $\mathbf{f}_s$ is the body force. The total stress tensor for the poroelastic structure is given by:

$$\boldsymbol{\sigma}_p = 2v_s D(\boldsymbol{\eta}) + \lambda (\nabla \cdot \boldsymbol{\eta}) \mathbf{I} - \alpha p_p \mathbf{I},$$

where v_s and λ denote the Lamé constants for the skeleton. The density of saturated medium is denoted by ρ_s, and the hydraulic conductivity is denoted by κ. In general, κ is a symmetric positive definite tensor, but in this work we assume an isotropic porous material so that κ is a scalar quantity. The constrained specific storage coefficient is denoted by s_0 and the Biot-Willis constant by α, which is usually close to unity. In the subsequent discussion, all the physical parameters are assumed to be constant in space and time. Note that the Biot system consists of the momentum equation for the balance of total forces (2.2a) and the mass conservation equation (2.2c), along with the standard assumption of Darcy's law (2.2b) for the flux.

We remark that model (2.2) is the same used in, e.g., [9], but it is different from the model used in other references, such as [3, 5]. References [3, 5] focus on blood-vessel interaction and assume the artery wall is a saturated poroelastic medium. Since here the focus is more general, we preferred to use model (2.2).

In order to complete the Stokes-Biot model, (2.1)–(2.2), we provide the following boundary, initial and interface conditions:

– Boundary conditions:

$$\boldsymbol{\sigma}_f \mathbf{n}_f = -P_{in}(t) \qquad \qquad \text{on } \Gamma_{in}^f , \tag{2.3a}$$

$$\boldsymbol{\sigma}_f \mathbf{n}_f = 0 \qquad \qquad \text{on } \Gamma_{out}^f , \tag{2.3b}$$

$$\mathbf{u}_f = 0 \qquad \qquad \text{on } \Gamma_t^f , \tag{2.3c}$$

$$\mathbf{u}_p \cdot \mathbf{n}_p = 0, \ \boldsymbol{\eta} = 0 \qquad \qquad \text{on } \Gamma_s^p , \tag{2.3d}$$

$$\mathbf{u}_p \cdot \mathbf{n}_p = 0, \ \boldsymbol{\sigma}_p \mathbf{n}_p = 0 \qquad \qquad \text{on } \Gamma_b^p . \tag{2.3e}$$

– Initial conditions:

$$\text{At } t = 0: \quad \mathbf{u}_f = \mathbf{0}, \quad p_p = 0, \quad \boldsymbol{\eta} = \mathbf{0}, \quad \boldsymbol{\eta}_t = \mathbf{0}. \tag{2.4}$$

– Interface conditions on Γ_I:

$$\mathbf{u}_f \cdot \mathbf{n}_f = -(\boldsymbol{\eta}_t + \mathbf{u}_p) \cdot \mathbf{n}_p, \tag{2.5a}$$

$$\boldsymbol{\sigma}_f \mathbf{n}_f = -\boldsymbol{\sigma}_p \mathbf{n}_p, \tag{2.5b}$$

$$\mathbf{n}_f \cdot \boldsymbol{\sigma}_f \mathbf{n}_f = -p_p, \tag{2.5c}$$

$$\mathbf{n}_f \cdot \boldsymbol{\sigma}_f \mathbf{t} = -\beta (\mathbf{u}_f - \boldsymbol{\eta}_t) \cdot \mathbf{t}, \tag{2.5d}$$

where $\mathbf{n}_f$ and $\mathbf{n}_p$ denote outward unit normal vectors to Ω^f and Ω^p, respectively, $\mathbf{t}$ denotes a unit tangent vector on Γ_I, and β denotes the resistance parameter in the tangential direction.

Here, (2.5a) describes the admissibility constraint. The conservation of momentum, expressed by (2.5b), requires that the total stress of the porous medium be balanced by the total stress of the fluid. For the balance of normal components of the stress in the fluid phase across the interface, we have (2.5c). Finally, the tangential stress of the fluid is assumed to be proportional to the slip rate according to the Beavers-Joseph-Saffman condition (2.5d). These interface conditions suffice to precisely couple the Stokes system (2.1) to the Biot system (2.2).

3 Semi-discrete Weak Formulation

Standard notation for Sobolev spaces and their associated norms and seminorms will be used to define a weak formulation of the problem. For example, $W^{m,p}(\Theta)$ is the usual Sobolev space with the norm $\| \cdot \|_{m,p,\Theta}$. In case of $p = 2$, the Sobolev space $W^{m,2}(\Theta)$ is denoted by $H^m(\Theta)$ with the norm $\| \cdot \|_{m,\Theta}$. When $m = 0$, $H^m(\Theta)$ coincides with $L^2(\Theta)$. In this case, the inner product and the norm will be denoted by $(\cdot, \cdot)_\Theta$ and $\| \cdot \|_\Theta$, respectively. Moreover, if $\Theta = \Omega^f$ or Ω^p, and the context is clear, Θ will be omitted, i.e., $(\cdot, \cdot) = (\cdot, \cdot)_{\Omega^f}$ or $(\cdot, \cdot)_{\Omega^p}$ for functions defined in Ω^f and Ω^p. For $\gamma \subset \partial \Omega^f \cup \partial \Omega^p$, we use $\langle \cdot, \cdot \rangle_\gamma$ to denote the duality pairing between $H^{-1/2}(\gamma)$ and $H^{1/2}(\gamma)$. Finally, the associated space of vector valued functions will be denoted by a boldface font.

Now, we are in a position to define the following function spaces for the velocities $\mathbf{u}_f$, $\mathbf{u}_p$, the pressures p_f, p_p, and the displacement $\boldsymbol{\eta}$, respectively:

$\mathbf{U}_f := \{\mathbf{v} \in \mathbf{H}^1(\Omega^f) : \mathbf{v} = \mathbf{0} \text{ on } \Gamma_t^f\},$

$\mathbf{U}_p := \mathbf{H}_{0,\Gamma_s^p \cup \Gamma_b^p}^{div}(\Omega^p) = \{\mathbf{v} \in \mathbf{L}^2(\Omega^p) : \nabla \cdot \mathbf{v} \in L^2(\Omega^p), \ \mathbf{v} \cdot \mathbf{n}_p = 0 \text{ on } \Gamma_s^p \cup \Gamma_b^p\},$

$Q_f := L^2(\Omega^f),$

$Q_p := L^2(\Omega^p),$

$\Sigma := \{\boldsymbol{\xi} \in \mathbf{H}^1(\Omega^p) : \boldsymbol{\xi} = \mathbf{0} \text{ on } \Gamma_s^p\}.$

We also define

$$\mathbf{G} := \mathbf{H}^{1/2}(\Gamma_I)$$

for the space of the control function to be introduced later.

Multiplying the governing equations (2.1) and (2.2) by appropriate test functions and using integration by parts, we obtain a continuous variational formulation for the fluid problem:

$$\rho_f \left(\frac{\partial \mathbf{u}_f}{\partial t}, \mathbf{v}_f \right) + 2\nu_f(D(\mathbf{u}_f), D(\mathbf{v}_f)) - (p_f, \nabla \cdot \mathbf{v}_f)$$

$$= (\mathbf{f}_f, \mathbf{v}_f) + \langle -P_{in}, \mathbf{v}_f \rangle_{\Gamma_{in}^f} + \langle \boldsymbol{\sigma}_f \mathbf{n}_f, \mathbf{v}_f \rangle_{\Gamma_I}, \quad \forall \mathbf{v}_f \in \mathbf{U}_f, \quad (3.1a)$$

$$(q_f, \nabla \cdot \mathbf{u}_f) = 0, \quad \forall q_f \in Q_f, \quad (3.1b)$$

and for the structure problem:

$$\rho_s \left(\frac{\partial^2 \boldsymbol{\eta}}{\partial t^2}, \boldsymbol{\xi} \right) + 2\nu_s(D(\boldsymbol{\eta}), D(\boldsymbol{\xi})) + \lambda(\nabla \cdot \boldsymbol{\eta}, \nabla \cdot \boldsymbol{\xi}) - \alpha(p_p, \nabla \cdot \boldsymbol{\xi})$$

$$= (\mathbf{f}_s, \boldsymbol{\xi}) + \langle \boldsymbol{\sigma}_p \mathbf{n}_p, \boldsymbol{\xi} \rangle_{\Gamma_I}, \quad \forall \boldsymbol{\xi} \in \Sigma, \quad (3.2a)$$

$$\kappa^{-1}(\mathbf{u}_p, \mathbf{v}_p) - (p_p, \nabla \cdot \mathbf{v}_p) = \langle -p_p, \mathbf{v}_p \cdot \mathbf{n}_p \rangle_{\Gamma_I}, \quad \forall \mathbf{v}_p \in \mathbf{U}_p, \quad (3.2b)$$

$$\left(q_p, \frac{\partial}{\partial t}(s_0 p_p + \alpha \nabla \cdot \boldsymbol{\eta}) + \nabla \cdot \mathbf{u}_p \right) = (q_p, f_p), \quad \forall q_p \in Q_p. \quad (3.2c)$$

Before we discretize the above equations in time, we introduce some notation first. Let $\Delta t = T/N$, where N is a positive integer and let $t^n = n\Delta t$. For any sufficiently smooth function $v(t, \mathbf{x})$, both constant and vector-valued, we define $v^n(\mathbf{x}) = v(t^n, \mathbf{x})$.

For the time discretization of the Stokes problem (3.1), we use the Backward Euler scheme. Then, the discrete-in-time, continuous-in-space problem of the free fluid reads as follows: For $n = 1, 2, \cdots, N$, find $\mathbf{u}_f^n \in \mathbf{U}_f$ and $p_f^n \in Q_f$ such that

$$\rho_f \left(\frac{\mathbf{u}_f^n - \mathbf{u}_f^{n-1}}{\Delta t}, \mathbf{v}_f \right) + 2\nu_f(D(\mathbf{u}_f^n), D(\mathbf{v}_f)) - (p_f^n, \nabla \cdot \mathbf{v}_f)$$

$$= (\mathbf{f}_f^n, \mathbf{v}_f) + \langle -P_{in}^n, \mathbf{v}_f \rangle_{\Gamma_{in}^f} + \langle \boldsymbol{\sigma}_f^n \mathbf{n}_f, \mathbf{v}_f \rangle_{\Gamma_I}, \quad \forall \mathbf{v}_f \in \mathbf{U}_f, \quad (3.3a)$$

$$(q_f, \nabla \cdot \mathbf{u}_f^n) = 0, \quad \forall q_f \in Q_f. \quad (3.3b)$$

On the other hand, the semi-discrete problem of the Biot model is: For $n = 1, 2, \cdots, N$, find $\boldsymbol{\eta}^n \in \Sigma$, $\mathbf{u}_p^n \in \mathbf{U}_p$, and $p_p^n \in Q_p$ such that

$$\rho_s \left(\frac{\boldsymbol{\eta}^n - 2\boldsymbol{\eta}^{n-1} + \boldsymbol{\eta}^{n-2}}{\Delta t^2}, \boldsymbol{\xi} \right) + 2\nu_s(D(\boldsymbol{\eta}^n), D(\boldsymbol{\xi})) + \lambda(\nabla \cdot \boldsymbol{\eta}^n, \nabla \cdot \boldsymbol{\xi})$$

$$- \alpha(p_p^n, \nabla \cdot \boldsymbol{\xi}) = (\mathbf{f}_s^n, \boldsymbol{\xi}) + \langle \sigma_p^n \mathbf{n}_p, \boldsymbol{\xi} \rangle_{\Gamma_I}, \quad \forall \boldsymbol{\xi} \in \Sigma, \qquad (3.4a)$$

$$\kappa^{-1}(\mathbf{u}_p^n, \mathbf{v}_p) - (p_p^n, \nabla \cdot \mathbf{v}_p) = \langle -p_p^n, \mathbf{v}_p \cdot \mathbf{n}_p \rangle_{\Gamma_I}, \quad \forall \mathbf{v}_p \in \mathbf{U}_p, \qquad (3.4b)$$

$$\left(q_p, s_0 \frac{p_p^n - p_p^{n-1}}{\Delta t} + \alpha \frac{\nabla \cdot \boldsymbol{\eta}^n - \nabla \cdot \boldsymbol{\eta}^{n-1}}{\Delta t} + \nabla \cdot \mathbf{u}_p^n \right) = (q_p, f_p^n), \quad \forall q_p \in Q_p.$$

$$(3.4c)$$

The fully-coupled scheme simultaneously solves these two subproblems, (3.3) and (3.4), coupled through the interface conditions (2.5).

4 Decoupling Schemes

The goal of this section is to develop efficient decoupling schemes that allow us to independently solve each subproblem while ensuring convergence to an accurate solution.

Let $\mathbf{g}^n = (g_1^n, g_2^n)^T := (\sigma_f^n \mathbf{n}_f)_{|_{\Gamma_I}}$. Then, using the interface condition (2.5b), we can rewrite (3.3a) and (3.4a), respectively, as

$$\rho_f \left(\frac{\mathbf{u}_f^n - \mathbf{u}_f^{n-1}}{\Delta t}, \mathbf{v}_f \right) + 2\nu_f(D(\mathbf{u}_f^n), D(\mathbf{v}_f)) - (p_f^n, \nabla \cdot \mathbf{v}_f)$$

$$= (\mathbf{f}_f^n, \mathbf{v}_f) + \langle -P_{in}, \mathbf{v}_f \rangle_{\Gamma_{in}^f} + \langle \mathbf{g}^n, \mathbf{v}_f \rangle_{\Gamma_I}, \quad \forall \mathbf{v}_f \in \mathbf{U}_f,$$

$$\rho_s \left(\frac{\boldsymbol{\eta}^n - 2\boldsymbol{\eta}^{n-1} + \boldsymbol{\eta}^{n-2}}{\Delta t^2}, \boldsymbol{\xi} \right) + 2\nu_s(D(\boldsymbol{\eta}^n), D(\boldsymbol{\xi})) + \lambda(\nabla \cdot \boldsymbol{\eta}^n, \nabla \cdot \boldsymbol{\xi})$$

$$- \alpha(p_p^n, \nabla \cdot \boldsymbol{\xi}) = (\mathbf{f}_s^n, \boldsymbol{\xi}) - \langle \mathbf{g}^n, \boldsymbol{\xi} \rangle_{\Gamma_I}. \quad \forall \boldsymbol{\xi} \in \Sigma.$$

On the other hand, we can rewrite

$$\mathbf{g}^n = \left((\mathbf{n}_f \cdot \sigma_f^n \mathbf{n}_f)\mathbf{n}_f + (\mathbf{t} \cdot \sigma_f^n \mathbf{n}_f)\mathbf{t} \right)|_{\Gamma_I}, \qquad (4.1)$$

which, together with (2.5c), implies that

$$-p_p|_{\Gamma_I} = (\mathbf{n}_f \cdot \sigma_f^n \mathbf{n}_f)|_{\Gamma_I} = \mathbf{g}^n \cdot \mathbf{n}_f = -\mathbf{g}^n \cdot \mathbf{n}_p.$$

Then, (3.4b) can be rewritten as

$$\kappa^{-1}(\mathbf{u}_p^n, \mathbf{v}_p) - (p_p^n, \nabla \cdot \mathbf{v}_p) = -\langle \mathbf{g}^n \cdot \mathbf{n}_p, \mathbf{v}_p \cdot \mathbf{n}_p \rangle_{\Gamma_I}, \quad \forall \mathbf{v}_p \in \mathbf{U}_p .$$

In summary, the semi-discrete Stokes and Biot problems, (3.3) and (3.4), can be rewritten in terms of $\mathbf{g}^n$:

$$\rho_f \left(\frac{\mathbf{u}_f^n - \mathbf{u}_f^{n-1}}{\Delta t}, \mathbf{v}_f \right) + 2\nu_f (D(\mathbf{u}_f^n), D(\mathbf{v}_f)) - (p_f^n, \nabla \cdot \mathbf{v}_f)$$

$$= (\mathbf{f}_f^n, \mathbf{v}_f) + \langle -P_{in}^n, \mathbf{v}_f \rangle_{\Gamma_{in}^f} + \langle \mathbf{g}^n, \mathbf{v}_f \rangle_{\Gamma_I}, \quad \forall \mathbf{v}_f \in \mathbf{U}_f , \tag{4.2a}$$

$$(q_f, \nabla \cdot \mathbf{u}_f^n) = 0, \quad \forall q_f \in Q_f . \tag{4.2b}$$

and

$$\rho_s \left(\frac{\eta^n - 2\eta^{n-1} + \eta^{n-2}}{\Delta t^2}, \boldsymbol{\xi} \right) + 2\nu_s (D(\eta^n), D(\boldsymbol{\xi})) + \lambda (\nabla \cdot \eta^n, \nabla \cdot \boldsymbol{\xi})$$

$$- \alpha (p_p^n, \nabla \cdot \boldsymbol{\xi}) = (\mathbf{f}_s^n, \boldsymbol{\xi}) - \langle \mathbf{g}^n, \boldsymbol{\xi} \rangle_{\Gamma_I}, \quad \forall \boldsymbol{\xi} \in \Sigma, \tag{4.3a}$$

$$\kappa^{-1}(\mathbf{u}_p^n, \mathbf{v}_p) - (p_p^n, \nabla \cdot \mathbf{v}_p) = -\langle \mathbf{g}^n \cdot \mathbf{n}_p, \mathbf{v}_p \cdot \mathbf{n}_p \rangle_{\Gamma_I}, \quad \forall \mathbf{v}_p \in \mathbf{U}_p , \tag{4.3b}$$

$$\left(q_p, s_0 \frac{p_p^n - p_p^{n-1}}{\Delta t} + \alpha \frac{\nabla \cdot \eta^n - \nabla \cdot \eta^{n-1}}{\Delta t} + \nabla \cdot \mathbf{u}_p^n \right) = (q_p, f_p^n), \quad \forall q_p \in Q_p. \tag{4.3c}$$

Note that these two subproblems are coupled through the function $\mathbf{g}^n$ only. If $\mathbf{g}^n$ is known at each time step n, then the two subproblems can be completely decoupled. However, $\mathbf{g}^n$ is unknown as σ_f^n is unknown.

Here, we will cast this fully-coupled problem as a constrained optimization problem using $\mathbf{g}^n$ as our control function. With an arbitrarily chosen $\mathbf{g}^n$, the solutions of (4.2) and (4.3) are not the same solutions for (3.3) and (3.4). It is because two interface conditions (2.5b) and (2.5c) are incorporated in the formulation, but the remaining interface conditions, (2.5a) and (2.5d), are not. Therefore, the objective of our optimization is to minimize the violation of (2.5a) and (2.5d). In order to do that, let the interface boundary Γ_I be partitioned into non-overlapping segments Γ_{I_i} for $i = 1, 2, \ldots, k$ such that $\Gamma_I = \cup_{i=1}^k \Gamma_{I_i}$. To satisfy the interface condition (2.5a) and (2.5d) at each time step n, we want to find a function $\mathbf{g}^n \in \mathbf{G}$ such that $(\mathbf{u}_f^n, p_f^n)$ satisfying (4.2) and $(\mathbf{u}_p^n, p_p^n, \eta^n)$ satisfying (4.3) minimize the functional

$$\mathscr{J}_n(\mathbf{g}^n) := \frac{1}{2} \sum_{i=1}^k \left(\frac{1}{\sqrt{|\Gamma_{I_i}|}} \int_{\Gamma_{I_i}} \mathbf{u}_f^n \cdot \mathbf{n}_f + \left(\frac{\eta^n - \eta^{n-1}}{\Delta t} + \mathbf{u}_p^n \right) \cdot \mathbf{n}_p \, d\Gamma_{I_i} \right)^2$$

$$+ \frac{1}{2} \left\| \mathbf{g}^n \cdot \mathbf{t} + \left(\beta \left(\mathbf{u}_f^n - \frac{\eta^n - \eta^{n-1}}{\Delta t} \right) \cdot \mathbf{t} \right) \right\|_{0,\Gamma_I}^2 + \frac{1}{2} \delta \|\mathbf{g}^n\|_{\mathbf{G}}^2, \tag{4.4}$$

where $|\gamma| := meas(\gamma)$ for $\gamma \subset \partial \Omega_f \cup \partial \Omega_p$ and $\delta > 0$ is a penalty parameter.

Minimizing the first term of the function in (4.4) seeks to weakly impose (2.5a) by forcing *flow balance* across the interface segments Γ_{I_i}. See [16] for details. The minimization of the second term in (4.4) is for the weak imposition of the Beavers-Joseph-Saffman condition (2.5d). Finally, the last term in (4.4) is a penalty term.

Remark 1 Thanks to (4.1), we can write $\mathbf{g}^n \cdot \mathbf{t}$ in place of $(\mathbf{n}_f \cdot \boldsymbol{\sigma}_f^n \mathbf{t})|_{\Gamma_I}$ in $\mathscr{J}_n(\mathbf{g}^n)$.

4.1 Least Squares Method

In this section, we are going to redefine the minimization problem as a least squares problem.

Set $\mathbf{F} = \mathbb{R}^k \times L^2(\Gamma_I) \times \mathbf{G}$ and define the operator $N_n : \mathbf{G} \to \mathbf{F}$ as

$$
N_n(\mathbf{g}^n) = \begin{pmatrix}
\frac{1}{\sqrt{|\Gamma_{I_1}|}} \int_{\Gamma_{I_1}} \mathbf{u}_f^n \cdot \mathbf{n}_f + \left(\frac{\eta^n - \eta^{n-1}}{\Delta t} + \mathbf{u}_p^n \right) \cdot \mathbf{n}_p \, d\Gamma_{I_1} \\
\vdots \\
\frac{1}{\sqrt{|\Gamma_{I_k}|}} \int_{\Gamma_{I_k}} \mathbf{u}_f^n \cdot \mathbf{n}_f + \left(\frac{\eta^n - \eta^{n-1}}{\Delta t} + \mathbf{u}_p^n \right) \cdot \mathbf{n}_p \, d\Gamma_{I_k} \\
\mathbf{g}^n \cdot \mathbf{t} + \left(\beta \left(\mathbf{u}_f^n - \frac{\eta^n - \eta^{n-1}}{\Delta t} \right) \cdot \mathbf{t} \right) \Big|_{\Gamma_I} \\
\sqrt{\delta} \, \mathbf{g}^n
\end{pmatrix}, \quad (4.5)
$$

where $(\mathbf{u}_f^n, p_f^n)$ is the solution of (4.2) and $(\mathbf{u}_p^n, p_p^n, \eta^n)$ is the solution of (4.3).

The minimization of the functional $\mathscr{J}_n(\mathbf{g}^n)$ in (4.4) is then equivalent to the minimization of the least squares function $\|N_n(\mathbf{g}^n)\|_{\mathbf{F}}^2$, that is:

$$
\min_{\mathbf{g}^n \in \mathbf{G}} \mathscr{J}_n(\mathbf{g}^n) = \frac{1}{2} \min_{\mathbf{g}^n \in \mathbf{G}} \|N_n(\mathbf{g}^n)\|_{\mathbf{F}}^2 . \quad (4.6)
$$

We solve this problem by a residual updating technique. First, an initial guess for a minimizer, $\mathbf{g}_{(0)}^n$, is chosen and $N_n(\mathbf{g}_{(0)}^n)$ is computed. Since we expect that $N_n(\mathbf{g}^n) \approx [\mathbf{0} \ \sqrt{\delta}\mathbf{g}^n]^T$ for a sufficiently small δ at the minimizer, we take $N_n(\mathbf{g}_{(0)}^n) - [\mathbf{0} \ \sqrt{\delta}\mathbf{g}_{(0)}^n]^T$ as a residual and find a correction $\mathbf{h}^n$ for $\mathbf{g}_{(0)}^n$ such that

$$
\frac{1}{2} \left\| \left(N_n(\mathbf{g}_{(0)}^n) - [\mathbf{0} \ \sqrt{\delta}\mathbf{g}_{(0)}^n]^T \right) + N_n'(\mathbf{g}_{(0)}^n)(\mathbf{h}^n) \right\|_{\mathbf{F}}^2
$$
$$
= \min_{\mathbf{y} \in \mathbf{G}} \frac{1}{2} \| \left(N_n(\mathbf{g}_{(0)}^n) - [\mathbf{0} \ \sqrt{\delta}\mathbf{g}_{(0)}^n]^T \right) + N_n'(\mathbf{g}_{(0)}^n)(\mathbf{y}) \|_{\mathbf{F}}^2 . \quad (4.7)
$$

Here, $N_n'(\mathbf{g}_{(0)}^n)(\cdot) : \mathbf{G} \to \mathbf{F}$ is defined by

$$N_n'(\mathbf{g}_{(0)}^n)(\mathbf{h}^n) = \begin{pmatrix} \frac{1}{\sqrt{|\Gamma_{I_1}|}} \int_{\Gamma_{I_1}} \mathbf{w}_f^n \cdot \mathbf{n}_f + \left(\frac{\boldsymbol{\varphi}^n}{\Delta t} + \mathbf{w}_p^n\right) \cdot \mathbf{n}_p \, d\Gamma_{I_1} \\ \vdots \\ \frac{1}{\sqrt{|\Gamma_{I_k}|}} \int_{\Gamma_{I_k}} \mathbf{w}_f^n \cdot \mathbf{n}_f + \left(\frac{\boldsymbol{\varphi}^n}{\Delta t} + \mathbf{w}_p^n\right) \cdot \mathbf{n}_p \, d\Gamma_{I_k} \\ \mathbf{h}^n \cdot \mathbf{t} + \left(\beta \left(\mathbf{w}_f^n - \frac{\boldsymbol{\varphi}^n}{\Delta t}\right) \cdot \mathbf{t}\right)\Big|_{\Gamma_I} \\ \sqrt{\delta}\mathbf{h}^n \end{pmatrix}, \qquad (4.8)$$

where $(\mathbf{w}_f^n, \phi_f^n)$ is the solution of the problem:

$$\rho_f \left(\frac{\mathbf{w}_f^n}{\Delta t}, \mathbf{v}_f\right) + 2\nu_f(D(\mathbf{w}_f^n), D(\mathbf{v}_f)) - (\phi_f^n, \nabla \cdot \mathbf{v}_f) = \langle \mathbf{h}^n, \mathbf{v}_f\rangle_{\Gamma_I}, \quad \forall \mathbf{v}_f \in \mathbf{U}_f, \tag{4.9a}$$

$$(q_f, \nabla \cdot \mathbf{w}_f^n) = 0, \quad \forall q_f \in Q_f, \tag{4.9b}$$

and $(\mathbf{w}_p^n, \phi_p^n, \boldsymbol{\varphi}^n)$ is the solution of the problem:

$$\rho_s \left(\frac{\boldsymbol{\varphi}^n}{\Delta t^2}, \boldsymbol{\xi}\right) + 2\nu_s(D(\boldsymbol{\varphi}^n), D(\boldsymbol{\xi})) + \lambda(\nabla \cdot \boldsymbol{\varphi}^n, \nabla \cdot \boldsymbol{\xi})$$

$$-\alpha(\phi_p^n, \nabla \cdot \boldsymbol{\xi}) = -\langle \mathbf{h}^n, \boldsymbol{\xi}\rangle_{\Gamma_I}, \quad \forall \boldsymbol{\xi} \in \Sigma, \tag{4.10a}$$

$$\kappa^{-1}(\mathbf{w}_p^n, \mathbf{v}_p) - (\phi_p^n, \nabla \cdot \mathbf{v}_p) = -\langle \mathbf{h}^n \cdot \mathbf{n}_p, \mathbf{v}_p \cdot \mathbf{n}_p\rangle_{\Gamma_I}, \quad \forall \mathbf{v}_p \in \mathbf{U}_p, \tag{4.10b}$$

$$\left(q_p, s_0 \frac{\phi_p^n}{\Delta t} + \alpha \frac{\nabla \cdot \boldsymbol{\varphi}^n}{\Delta t} + \nabla \cdot \mathbf{w}_p^n\right) = 0, \quad \forall q_p \in Q_p. \tag{4.10c}$$

In order to solve the minimization problem (4.7), we solve its normal equation

$$N_n'(\mathbf{g}_{(0)}^n)^* N_n'(\mathbf{g}_{(0)}^n)(\mathbf{h}^n) = -N_n'(\mathbf{g}_{(0)}^n)^* \left(N_n(\mathbf{g}_{(0)}^n) - [\mathbf{0} \ \sqrt{\delta}\mathbf{g}_{(0)}^n]^T\right), \tag{4.11}$$

where $N_n'(\mathbf{g}_{(0)}^n)^* : \mathbb{R}^k \times L^2(\Gamma_I) \times \mathbf{G}^* \to \mathbf{G}^*$ is the adjoint operator of $N_n'(\mathbf{g}_{(0)}^n)$ identified in the following lemma.

Lemma 1 *For* $(\gamma, y, \mathbf{z}) \in \mathbb{R}^k \times L^2(\Gamma_I) \times \mathbf{G}^*$, *the adjoint of* $N_n'(\mathbf{g}_{(0)}^n)$ *is given by*

$$N_n'(\mathbf{g}_{(0)}^n)^* \begin{pmatrix} \gamma \\ y \\ \mathbf{z} \end{pmatrix} = \left(\bar{\mathbf{w}}_f^n - \bar{\boldsymbol{\varphi}}^n - (\bar{\mathbf{w}}_p^n \cdot \mathbf{n}_p)\mathbf{n}_p\right)\Big|_{\Gamma_I} + y\mathbf{t} + \sqrt{\delta}\mathbf{z}, \tag{4.12}$$

where $(\bar{\mathbf{w}}_f^n, \bar{\phi}_f^n)$ is the solution of

$$\rho_f \left(\frac{\bar{\mathbf{w}}_f^n}{\Delta t}, \mathbf{v}_f \right) + 2\nu_f (D(\bar{\mathbf{w}}_f^n), D(\mathbf{v}_f)) - (\bar{\phi}_f^n, \nabla \cdot \mathbf{v}_f)$$

$$= \beta \langle y, \mathbf{v}_f \cdot \mathbf{t} \rangle_{\Gamma_I} + \sum_{i=1}^{k} \gamma_i \frac{1}{\sqrt{|\Gamma_{I_i}|}} \int_{\Gamma_{I_i}} \mathbf{v}_f \cdot \mathbf{n}_f \, d\Gamma_{I_i}, \ \forall \mathbf{v}_f \in \mathbf{U}_f,$$

$$\text{(4.13a)}$$

$$(q_f, \nabla \cdot \bar{\mathbf{w}}_f^n) = 0, \quad \forall q_f \in Q_f, \tag{4.13b}$$

and $(\bar{\mathbf{w}}_p^n, \bar{\phi}_p^n, \bar{\boldsymbol{\varphi}}^n)$ is the solution of

$$\rho_s \left(\frac{\bar{\boldsymbol{\varphi}}^n}{\Delta t^2}, \boldsymbol{\xi} \right) + 2\nu_s (D(\bar{\boldsymbol{\varphi}}^n), D(\boldsymbol{\xi})) + \lambda (\nabla \cdot \bar{\boldsymbol{\varphi}}^n, \nabla \cdot \boldsymbol{\xi}) + \frac{\alpha}{\Delta t} (\bar{\phi}_p^n, \nabla \cdot \boldsymbol{\xi})$$

$$= -\frac{\beta}{\Delta t} \langle y, \boldsymbol{\xi} \cdot \mathbf{t} \rangle_{\Gamma_I} + \frac{1}{\Delta t} \sum_{i=1}^{k} \gamma_i \frac{1}{\sqrt{|\Gamma_{I_i}|}} \int_{\Gamma_{I_i}} \boldsymbol{\xi} \cdot \mathbf{n}_p \, d\Gamma_{I_i}, \quad \forall \boldsymbol{\xi} \in \Sigma, \tag{4.14a}$$

$$\kappa^{-1}(\bar{\mathbf{w}}_p^n, \mathbf{v}_p) + (\bar{\phi}_p^n, \nabla \cdot \mathbf{v}_p) = \sum_{i=1}^{k} \gamma_i \frac{1}{\sqrt{|\Gamma_{I_i}|}} \int_{\Gamma_{I_i}} \mathbf{v}_p \cdot \mathbf{n}_p \, d\Gamma_{I_i}, \quad \forall \mathbf{v}_p \in \mathbf{U}_p, \tag{4.14b}$$

$$\left(q_p, s_0 \frac{\bar{\phi}_p^n}{\Delta t} - \alpha \nabla \cdot \bar{\boldsymbol{\varphi}}^n - \nabla \cdot \bar{\mathbf{w}}_p^n \right) = 0, \quad \forall q_p \in Q_p. \tag{4.14c}$$

Proof Taking $(\mathbf{v}_f, q_f) = (\bar{\mathbf{w}}_f^n, \bar{\phi}_f^n)$, $(\mathbf{v}_p, q_p, \boldsymbol{\xi}) = (\bar{\mathbf{w}}_p^n, \bar{\phi}_p^n, \bar{\boldsymbol{\varphi}}^n)$ in (4.9) and (4.10), respectively, and $(\mathbf{v}_f, q_f) = (\mathbf{w}_f^n, \phi_f^n)$, $(\mathbf{v}_p, q_p, \boldsymbol{\xi}) = (\mathbf{w}_p^n, \phi_p^n, \boldsymbol{\varphi}^n)$ in (4.13) and (4.14), respectively, we obtain

$$\langle \mathbf{h}^n, \bar{\mathbf{w}}_f^n - \bar{\boldsymbol{\varphi}}^n - (\bar{\mathbf{w}}_p^n \cdot \mathbf{n}_p) \mathbf{n}_p \rangle_{\Gamma_I} = \sum_{i=1}^{k} \gamma_i \frac{1}{\sqrt{|\Gamma_{I_i}|}} \int_{\Gamma_{I_i}} \mathbf{w}_f^n \cdot \mathbf{n}_f + \left(\frac{\boldsymbol{\varphi}^n}{\Delta t} + \mathbf{w}_p^n \right) \cdot \mathbf{n}_p \, d\Gamma_{I_i}$$

$$+ \left\langle y, \beta \left(\mathbf{w}_f^n - \frac{\boldsymbol{\varphi}^n}{\Delta t} \right) \cdot \mathbf{t} \right\rangle_{\Gamma_I}. \tag{4.15}$$

Hence, by (4.8), (4.12), and (4.15), for $\mathbf{h}^n \in \mathbf{G}$ we have:

$$\left(N_n'(\mathbf{g}_{(0)}^n)(\mathbf{h}^n), \begin{bmatrix} \gamma \\ y \\ \mathbf{z} \end{bmatrix} \right) = \sum_{i=1}^{k} \gamma_i \frac{1}{\sqrt{|\Gamma_{I_i}|}} \int_{\Gamma_{I_i}} \mathbf{w}_f^n \cdot \mathbf{n}_f + \left(\frac{\boldsymbol{\varphi}^n}{\Delta t} + \mathbf{w}_p^n \right) \cdot \mathbf{n}_p \, d\Gamma_{I_i}$$

$$+ \left\langle y, \mathbf{h}^n \cdot \mathbf{t} + \beta \left(\mathbf{w}_f^n - \frac{\boldsymbol{\varphi}^n}{\Delta t} \right) \cdot \mathbf{t} \right\rangle_{\Gamma_I} + \sqrt{\delta} \langle \mathbf{h}^n, \mathbf{z} \rangle_{\Gamma_I}$$

$$= \langle \mathbf{h}^n, \bar{\mathbf{w}}_f^n - \bar{\boldsymbol{\varphi}}^n - (\bar{\mathbf{w}}_p^n \cdot \mathbf{n}_p) \mathbf{n}_p + y\mathbf{t} + \sqrt{\delta} \mathbf{z} \rangle_{\Gamma_I}$$

$$= \left(\mathbf{h}^n, N'_n(\mathbf{g}^n_{(0)})^* \left(\begin{bmatrix} \gamma \\ y \\ \mathbf{z} \end{bmatrix} \right) \right).$$

For the solution of (4.11), we use the Conjugate Gradient (CG) algorithm; see, e.g., [17]. The steps of the CG algorithm applied to the solution of problem $A^*A\mathbf{x} = A^*\mathbf{b}$ are described in Algorithm 1. Here, ε is a prescribed error tolerance. Note that the normal equation does not need to be formed explicitly for the algorithm.

Algorithm 1 Conjugate Gradient (CG) method for the least squares problem

---

1. Initialize $\mathbf{x}_{(0)}$.

2. Set $\mathbf{r}_{(0)} = \mathbf{b} - A\mathbf{x}_{(0)}$, $\mathbf{p}_{(0)} = A^*\mathbf{r}_{(0)}$.

3. For $i = 0, 1, 2, \cdots$,

    a. if $\|A^*\mathbf{r}_{(i)}\| < \varepsilon$, stop,
    b. $\sigma_{(i)} = \|A^*\mathbf{r}_{(i)}\|^2/\|A\mathbf{p}_{(i)}\|^2$,
    c. $\mathbf{x}_{(i+1)} = \mathbf{x}_{(i)} + \sigma_{(i)}\mathbf{p}_{(i)}$,
    d. $\mathbf{r}_{(i+1)} = \mathbf{r}_{(i)} - \sigma_{(i)}A\mathbf{p}_{(i)}$,
    e. $\tau_{(i)} = \|A^*\mathbf{r}_{(i+1)}\|^2/\|A^*\mathbf{r}_{(i)}\|^2$,
    f. $\mathbf{p}_{(i+1)} = A^*\mathbf{r}_{(i+1)} + \tau_{(i)}\mathbf{p}_{(i)}$.

Once $\mathbf{h}^n$ has been computed, the least squares problem (4.6) can be solved using the residual updating algorithm described in Algorithm 2.

Algorithm 2 Residual updating algorithm

---

1. Initialize $\mathbf{g}^n_{(0)}$,

2. Solve Stokes/Biot problem defined by (4.2) and (4.3) to get $\mathbf{u}^n_f, p^n_f, \mathbf{u}^n_p, p^n_p$, and $\boldsymbol{\eta}^n$,

3. Compute $N(\mathbf{g}^n_{(0)})$,

4. Find the correction $\mathbf{h}^n$ using the CG algorithm (Algorithm 1) with $A = N'_n(\mathbf{g}^n_{(0)})$, $\mathbf{b} = -(N_n(\mathbf{g}^n_{(0)}) - [0 \ \sqrt{\delta}\mathbf{g}^n_{(0)}]^T)$, $\mathbf{x} = \mathbf{h}^n$,

5. $\mathbf{g}^n \leftarrow \mathbf{g}^n_{(0)} + \mathbf{h}^n$.

4.2 Linear Equation

In this section, we suppose that $\mathbf{u}^n_p$ is regular enough that $\mathbf{u}^n_p \cdot \mathbf{n}_p \in L^2(\Gamma_I)$. In this case, the objective functional $\mathscr{J}_n$ can be defined as:

$$\mathscr{J}_n(\mathbf{g}^n) := \frac{1}{2} \left\| \left(\mathbf{u}_f^n \cdot \mathbf{n}_f + \left(\frac{\eta^n - \eta^{n-1}}{\Delta t} + \mathbf{u}_p^n \right) \cdot \mathbf{n}_p \right) \Big|_{\Gamma_I} \right\|_{L^2(\Gamma_I)}^2$$

$$+ \frac{1}{2} \left\| \mathbf{g}^n \cdot \mathbf{t} + \beta \left(\left(\mathbf{u}_f^n - \frac{\eta^n - \eta^{n-1}}{\Delta t} \right) \cdot \mathbf{t} \right) \Big|_{\Gamma_I} \right\|_{L^2(\Gamma_I)}^2$$

$$+ \frac{1}{2} \delta \|\mathbf{g}^n\|_{\mathbf{G}}^2. \tag{4.16}$$

Assuming no penalty term in (4.16) and choosing a control space $\mathbf{G} := \mathbf{L}^2(\Gamma_I)$, define the linear operator $L_n : \mathbf{G} \to \mathbf{G}$ by

$$L_n(\mathbf{g}^n) = \begin{pmatrix} \left(\mathbf{u}_f^n \cdot \mathbf{n}_f + \left(\frac{\eta^n - \eta^{n-1}}{\Delta t} + \mathbf{u}_p^n \right) \cdot \mathbf{n}_p \right) \Big|_{\Gamma_I} \\ \mathbf{g}^n \cdot \mathbf{t} + \left(\beta \left(\mathbf{u}_f^n - \frac{\eta^n - \eta^{n-1}}{\Delta t} \right) \cdot \mathbf{t} \right) \Big|_{\Gamma_I} \end{pmatrix}, \tag{4.17}$$

where $(\mathbf{u}_f^n, p_f^n)$ satisfies (4.2) and $(\mathbf{u}_p^n, p_p^n, \eta^n)$ satisfies (4.3). Assuming further that the unknown stress $\mathbf{g}^n$ and unknowns $\mathbf{u}_f^n$, $\mathbf{u}_p^n$, η^n have the same number of degrees of freedom on the interface when discretized (this is easily achieved by using a fluid mesh and a structure mesh that match at the interface), we can convert the minimization problem to the following linear problem:

$$\text{Find } \mathbf{g}^n \in \mathbf{G} \text{ such that } L_n(\mathbf{g}^n) = \mathbf{0}. \tag{4.18}$$

We can solve (4.18) using a residual updating technique described in Algorithm 4: For a given initial guess $\mathbf{g}_{(0)}^n$, find $\mathbf{h}^n$ such that

$$L_n(\mathbf{g}_{(0)}^n) + L_n'(\mathbf{g}_{(0)}^n)(\mathbf{h}^n) = \mathbf{0}$$

and update $\mathbf{g}^n$. Here, $L_n' : \mathbf{G} \to \mathbf{G}$ is defined by

$$L_n'(\mathbf{g}_{(0)}^n)(\mathbf{h}^n) = \begin{pmatrix} \left(\mathbf{w}_f^n \cdot \mathbf{n}_f + \left(\frac{\varphi^n}{\Delta t} + \mathbf{w}_p^n \right) \cdot \mathbf{n}_p \right) \Big|_{\Gamma_I} \\ \mathbf{h}^n \cdot \mathbf{t} + \left(\beta \left(\mathbf{w}_f^n - \frac{\varphi^n}{\Delta t} \right) \cdot \mathbf{t} \right) \Big|_{\Gamma_I} \end{pmatrix}, \tag{4.19}$$

where $(\mathbf{w}_f^n, \phi_f^n)$ is the solution to (4.9) and $(\mathbf{w}_p^n, \phi_p^n, \varphi^n)$ is the solution to (4.10).

Note that the operator L_n' is not self-adjoint. Therefore, the residual updating technique can be used in combination with an iterative solver for a non-self-adjoint problem such as BiCGSTAB method; see, e.g., [17]. The algorithm for the BiCGSTAB method is provided in Algorithm 3.

Algorithm 3 Biconjugate Gradient stabilized (BiCGSTAB) method
---

1. Initialize $\mathbf{x}_{(0)}$,
2. Set $\mathbf{r}_{(0)} = \mathbf{b} - A\mathbf{x}_{(0)}$,
3. Choose an arbitrary vector $\hat{\mathbf{r}}_{(0)}$ such that $(\hat{\mathbf{r}}_{(0)}, \mathbf{r}_{(0)}) \neq 0$, e.g., $\hat{\mathbf{r}}_{(0)} = \mathbf{r}_{(0)}$,
4. Set $\rho_{(0)} = \alpha = \omega_{(0)} = 1$,
5. Set $\mathbf{v}_{(0)} = \mathbf{p}_{(0)} = \mathbf{0}$,
6. For $i = 1, 2, \cdots$,

    a. if $\|\mathbf{r}_{(i-1)}\| < \varepsilon$ stop,
    b. $\rho_{(i)} = (\hat{\mathbf{r}}_{(0)}, \mathbf{r}_{(i-1)})$,
    c. $\beta_{(i)} = (\rho_{(i)}/\rho_{(i-1)})(\alpha_{(i)}/\omega_{(i-1)})$,
    d. $\mathbf{p}_{(i)} = \mathbf{r}_{(i-1)} + \beta_{(i)}(\mathbf{p}_{(i-1)} - \omega_{(i-1)}\mathbf{v}_{(i-1)})$,
    e. $\mathbf{v}_{(i)} = A\mathbf{p}_{(i)}$,
    f. $\alpha_{(i)} = \rho_{(i)}/(\hat{\mathbf{r}}_{(0)}, \mathbf{v}_{(i)})$,
    g. $\mathbf{s}_{(i)} = \mathbf{r}_{(i-1)} - \alpha_{(i)}\mathbf{v}_{(i)}$,
    h. $\mathbf{t}_{(i)} = A\mathbf{s}_{(i)}$,
    i. $\omega_{(i)} = (\mathbf{t}_{(i)}, \mathbf{s}_{(i)})/(\mathbf{t}_{(i)}, \mathbf{t}_{(i)})$
    j. $\mathbf{x}_{(i)} = \mathbf{x}_{(i-1)} + \alpha_{(i)}\mathbf{p}_{(i)} + \omega_{(i)}\mathbf{s}_{(i)}$,
    k. $\mathbf{r}_{(i)} = \mathbf{s}_{(i)} - \omega_{(i)}\mathbf{t}_{(i)}$.

Algorithm 4 Residual updating algorithm for the linear equation
---

1. Initialize $\mathbf{g}_{(0)}^n$.

2. Solve Stokes/Biot problem defined by (4.2) and (4.3) for $\mathbf{u}_f^n, p_f^n, \mathbf{u}_p^n, p_p^n, \boldsymbol{\eta}^n$.

3. Compute $L(\mathbf{g}_{(0)}^n)$.

4. Find the correction $\mathbf{h}^n$ using the BiCGSTAB algorithm (Algorithm 3) with $A = L_n'(\mathbf{g}_{(0)}^n)$, $\mathbf{b} = -L_n(\mathbf{g}_{(0)}^n)$, and $\mathbf{x} = \mathbf{h}^n$.

5. $\mathbf{g}^n \leftarrow \mathbf{g}_{(0)}^n + \mathbf{h}^n$.

---

5 Numerical Experiments

In order to investigate the convergence properties of Algorithms 2 and 4, we performed numerical experiments using a non-physical example.

We take $\Omega_p = (0, 1) \times (0, 1)$ for the poroelastic structure. The fluid domain $\Omega_f = (0, 1) \times (1, 2)$ is superposed on Ω_p, with the fluid-structure interface $\Gamma_I = \{(x, y) : 0 < x < 1, y = 1\}$. Also, the physical parameters are chosen as follows: $\nu_f = \nu_s = 0.5$, $\rho_f = \rho_s = 1$, $\alpha = \beta = \lambda = s_0 = \kappa = 1$. The right-hand side functions $\mathbf{f}_f$, $\mathbf{f}_s$, and f_p are chosen so that the exact solution is:

$$\mathbf{u}_f = [(y-1)^2 x^3 (1+t^2) , -\cos(y)e(1+t^2)],$$
$$p_f = (\cos(x)e^y + y^2 - 2y + 1)(1+t^2),$$
$$\mathbf{u}_p = [-x(\sin(y)e + 2(y-1))(1+t^2) , (-\cos(y)e + (y-1)^2)(1+t^2)],$$
$$p_p = (-\sin(y)e + \cos(x)e^y + y^2 - 2y + 1)(1+t^2),$$
$$\eta = [\sqrt{2}\cos(\sqrt{2}x)\cos(y)(1+t^2) , \sin(\sqrt{2}x)\sin(y)(1+t^2)].$$

The boundary and initial conditions are determined using the exact solution.

Figure 2 shows the magnitude of the fluid velocity and the fluid pressure at time $t = 0.0005$, while Figure 3 displays the magnitude of the structure displacement, the magnitude of the Darcy velocity, and the structure pressure at the same time.

Fig. 2 Exact solution for the fluid problem at time $t = 0.0005$: (a) the magnitude of the fluid velocity and (b) the fluid pressure.

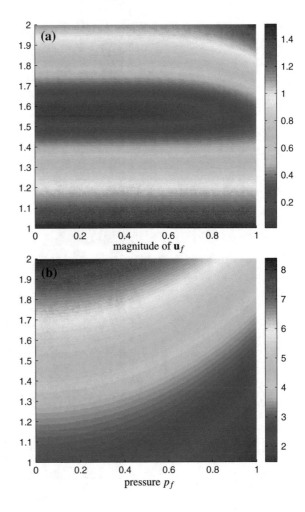

Fig. 3 Exact solution for the structure problem at time $t = 0.0005$: (a) the magnitude of the structure displacement, (b) the magnitude of the Darcy velocity, and (c) the fluid pressure.

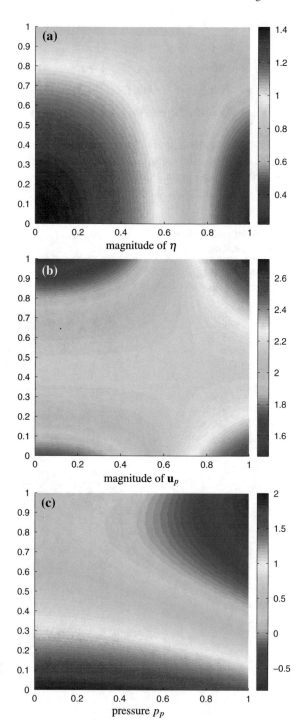

Note that the chosen exact solution does not satisfy all the interface conditions. Instead, it satisfies only (2.5b) and (2.5c), but not (2.5a) and (2.5d). Indeed, the exact solution satisfies the following variations of (2.5a) and (2.5d):

$$\mathbf{u}_f \cdot \mathbf{n}_f = -\mathbf{u}_p \cdot \mathbf{n}_p, \tag{5.1a}$$

$$\mathbf{n}_f \cdot \sigma_f \mathbf{t} = -\beta(\mathbf{u}_f \cdot \mathbf{t}). \tag{5.1b}$$

For our numerical implementation, we still implemented our algorithms as if all the interface conditions, (2.5a)–(2.5d), are satisfied. Then, to compensate the inexact interface conditions, we modified the functionals $N_n(\mathbf{g}^n)$ and $L_n(\mathbf{g}^n)$. More specifically, we compute the functional (4.5) with the additional term $-\eta_t \cdot \mathbf{n}$ for the first k entries and $\beta \, \eta_t \cdot \mathbf{t}$ for entry $k + 1$, where η_t is given by the chosen exact solution. Similarly, $-\eta_t \cdot \mathbf{n}$ and $\beta \, \eta_t \cdot \mathbf{t}$ are added in the first and second entries of (4.17), respectively.

For the finite element approximations, we used inf-sup stable Taylor-Hood elements $\mathbb{P}_2 - \mathbb{P}_1$ on structured meshes for both $(\mathbf{u}_f, p_f)$ and $(\mathbf{u}_p, p_p)$, and $\mathbb{P}_2$ elements for η. Because these elements are not stable for the Biot model, we added a stabilization term $\mu(\nabla \cdot \mathbf{u}_p, \nabla \cdot \mathbf{v}_p)$ to the Darcy equation (4.3) with $\mu = 10$ and analogous terms were added to (4.10b) and (4.14b).

In order to verify the convergence of Algorithms 2 and 4, numerical experiments were performed with varying mesh sizes. The time step size Δt was set to 0.0001. The error tolerance for both CG and BiCGSTAB methods used in Algorithm 2 and Algorithm 4, respectively, was set to $\varepsilon = 10^{-5}$. We started with $\mathbf{g}_{(0)}^1 = -(0.1, 0.1)^T$ and $\mathbf{h}_{(0)}^1 = (0.01, 0.01)^T$ for the first time step. After the first time step, the initial stress function was chosen as $\mathbf{g}_{(0)}^n = \mathbf{g}^{n-1}$ for both algorithms.

First, we investigated Algorithm 2 through a mesh refinement study; we halved the mesh size in each mesh refinement. Table 1 reports the errors of finite element solutions with four different meshes at the fifth time step ($t = 0.0005$), where $(\mathbf{u}_f^h, p_f^h, \mathbf{u}_p^h, p_p^h, \eta^h)$, $(\mathbf{u}_f^{ex}, p_f^{ex}, \mathbf{u}_p^{ex}, p_p^{ex}, \eta^{ex})$ denote the finite element solution and the

Table 1 Errors and convergence rates using Algorithm 2 at the fifth time step ($t = 0.0005$) with $\Delta t = 0.0001$.

h	$\|\mathbf{u}_f^h - \mathbf{u}_f^{ex}\|_{0,\Omega_f}$	rate	$|\mathbf{u}_f^h - \mathbf{u}_f^{ex}|_{1,\Omega_f}$	rate	$\|p_f^h - p_f^{ex}\|_{0,\Omega_f}$	rate
1/2	1.18×10^{-2}	–	1.59×10^{-1}	–	1.38×10^{-1}	–
1/4	1.26×10^{-3}	3.23	3.59×10^{-2}	2.15	2.51×10^{-2}	2.46
1/8	1.43×10^{-4}	3.14	8.69×10^{-3}	2.05	5.25×10^{-3}	2.26
1/16	1.75×10^{-5}	3.03	2.15×10^{-3}	2.02	1.21×10^{-3}	2.12
h	$|\mathbf{u}_p^h - \mathbf{u}_p^{ex}|_{H^{div}(\Omega_p)}$	rate	$\|p_p^h - p_p^{ex}\|_{0,\Omega_p}$	rate	$|\eta^h - \eta^{ex}|_{1,\Omega_p}$	rate
1/2	7.12×10^{-3}	–	4.58×10^{-2}	–	6.33×10^{-2}	–
1/4	1.93×10^{-3}	1.88	1.12×10^{-2}	2.03	1.58×10^{-2}	2.00
1/8	4.72×10^{-4}	2.03	2.78×10^{-3}	2.01	3.95×10^{-3}	2.00
1/16	1.16×10^{-4}	2.02	6.95×10^{-4}	2.00	9.87×10^{-4}	2.00

Table 2 Number of CG iterations for Algorithm 2, initial and terminal functional values at the first time step ($t = 0.0001$) with $\Delta t = 0.0001$.

h	No. of CG iter.	Initial $\mathscr{J}_n$	Terminal $\mathscr{J}_n$
1/2	10	1.12×10^3	6.63×10^{-8}
1/4	26	7.54×10^2	8.49×10^{-7}
1/8	62	2.44×10^2	2.91×10^{-7}
1/16	161	7.25×10^1	9.49×10^{-8}

Table 3 Number of CG iterations for Algorithm 2, initial and terminal residual values for the fifth time step ($t = 0.0005$) with $\Delta t = 0.0001$.

h	No. of CG iter.	Initial $\mathscr{J}_n$	Terminal $\mathscr{J}_n$
1/2	9	2.58×10^{-5}	1.30×10^{-14}
1/4	21	3.64×10^{-5}	1.23×10^{-12}
1/8	57	5.16×10^{-6}	3.31×10^{-14}
1/16	94	2.85×10^{-7}	1.92×10^{-12}

exact solution, respectively. The rates of convergence of all the quantities in the respective norms are those predicted by the theory.

Table 2 and 3 report the number of CG iterations for Algorithm 2 together with the initial and final values of the objective functional $\mathscr{J}_n$ defined in (4.4) at the first time step and at the fifth time step, respectively. We remark that while at the first time step the number of CG iterations more than doubles every time the mesh size is halved, it is no more the case at the fifth time step when going from the third to the fourth mesh refinement. Recall that we make a random initial choice for $\mathbf{g}_{(0)}^1$. At every successive step, however, we set $\mathbf{g}_{(0)}^n = \mathbf{g}^{n-1}$. Therefore, the initial value of $\mathscr{J}_n$ is much larger at the first time step than at the fifth. Note that the initial value of $\mathscr{J}_n$ at the fifth time step is already around 10^{-5}.

Next, we investigated Algorithm 4. We observed that Algorithm 4 gave almost identical errors to the ones reported in Table 1, hence the report is omitted here. The number of BiCGSTAB iterations for Algorithm 4, and the initial and final values of $\mathscr{J}_n$ at the first time step and at the fifth time step are presented in Tables 4 and 5, respectively.

We observe that at the first time step, with an arbitrarily chosen $\mathbf{g}_{(0)}^1$, the number of CG iterations needed is similar to or a little more than that of BiCGSTAB iterations. However, at the fifth time step, where the iterations start with a more accurate initial control $\mathbf{g}_{(0)}^5 = \mathbf{g}^4$, significantly less iterations are needed for CG than BiCGSTAB iterations. This gets more prominent as h gets smaller.

Table 4 Number of BiCGSTAB iterations for Algorithm 4, initial and terminal functional values at the first time step ($t = 0.0001$) with $\Delta t = 0.0001$.

h	No. of BiCGSTAB iter.	Initial $\mathscr{J}_n$	Terminal $\mathscr{J}_n$
1/2	8	1.34×10^3	1.05×10^{-7}
1/4	23	8.69×10^2	1.16×10^{-6}
1/8	68	2.64×10^2	7.17×10^{-7}
1/16	132	7.52×10^1	6.77×10^{-7}

Table 5 Number of BiCGSTAB iterations for Algorithm 4, initial and terminal residual values for the fifth time step ($t = 0.0005$) with $\Delta t = 0.0001$.

h	No. of BiCGSTAB iter.	Initial $\mathscr{J}_n$	Terminal $\mathscr{J}_n$
1/2	7	1.20×10^{-3}	3.91×10^{-11}
1/4	34	1.70×10^{-4}	2.69×10^{-11}
1/8	121	1.34×10^{-5}	2.43×10^{-11}
1/16	263	4.94×10^{-7}	3.85×10^{-11}

As far as computational effort is concerned, the computational cost per iteration is almost the same for both algorithms. More specifically, in each CG iteration, (4.9)–(4.10) and (4.13)–(4.14) need to be solved, while in each BiCGSTAB iteration, (4.9)–(4.10) need to be solved twice with different right-hand sides.

6 Conclusions

We have studied the interaction of a free-fluid with a poroelastic structure, modeled by the Stokes-Biot system. After discussing the time-discretized variational formulation of the coupled problem, we developed a minimization problem in which the Stokes-Biot system was decoupled through a control function on the interface. Two numerical algorithms based on a residual updating technique have been proposed; one solves a least squares problem and the other solves a linear problem when the solution $\mathbf{u}_p$ is smooth enough. We observed that both algorithms yielded an optimal control function along with a solution for the Stokes-Biot system that satisfies the interface conditions. Also, the optimal rates of convergence for finite element solutions demonstrate that there was no spatial degradation of the solution over time steps. On the other hand, the proposed decoupling schemes enable us to solve the two subproblems in parallel and they can be easily extended to a nonlinear system, e.g., the coupled Navier-Stokes and Biot system. Subsequent work will provide an analytical framework for the proposed methods and include numerical experiments that are designed for moving domains in a physical setting.

Acknowledgments This work was supported by the Institute for Mathematics and its Applications (IMA), which is funded by the National Science Foundation (NSF). A. Cesmelioglu would like to thank Oakland University for the URC Faculty Research Fellowship Award. H. Lee was supported by NSF under contract number DMS 1418960 and A. Quaini's research was supported in part by NSF under grant DMS-1262385. The research of S.-Y. Yi was supported by NSF under contract number DMS 1217123.

References

1. M. A. Murad, J. N. Guerreiro, and A. F. D. Loula. Micromechanical computational modeling of reservoir compaction and surface subsidence. *Math. Contemp.*, 19:41–69, 2000.
2. M. A. Murad, J. N. Guerreiro, and A. F. D. Loula. Micromechanical computational modeling of secondary consolidation and hereditary creep in soils. *Comput. Methods Appl. Mech. Engrg.*, 190(15-17):1985–2016, 2001.
3. N. Koshiba, J. Ando, X. Chen, and T. Hisada. Multiphysics simulation of blood flow and LDL transport in a porohyperelastic arterial wall model. *J. of Biomech. Eng.*, 129:374–385, 2007.
4. V. M. Calo, N. F. Brasher, Y. Bazilevs, and T. J. R. Hughes. Multiphysics model for blood flow and drug transport with application to patient-specific coronary artery flow. *Comput. Mech.*, 43(1):161–177, 2008.
5. S. Badia, A. Quaini, A. Quateroni, Coupling Biot and Navier-Stokes equations for modeling fluid-poroelastic media interaction, Journal of Computational Physics, 228:7986–8014, 2009.
6. B. Tully and Y. Ventikos. Coupling poroelasticity and CFD for cerebrospinal fluid hydrodynamics. *Biomedical Engineering, IEEE Transactions on*, 56(6):1644-1651, 2009.
7. B. Ganis, R. Liu, B. Wang, M.F. Wheeler, and I. Yotov. Multiscale modeling of flow and geomechanics. *Radon Series on Computational and Applied Mathematics*, pages 165–204, 2013.
8. M. Bukac, I. Yotov, and P. Zunino, An operator splitting approach for the interaction between a fluid and a multilayered poroelastic structure. *Numerical Methods for Partial Differential Equations*, 31(4):1054–1100, 2015.
9. M. Bukač, I. Yotov, R. Zakerzadeh and P. Zunino, Partitioning strategies for the interaction of a fluid with a poroelastic material based on a Nitsches coupling approach, Computer Methods in Applied Mechanics and Engineering, 292:138–170, 1 August 2015.
10. M. A. Biot. General theory of three-dimensional consolidation. *J. Appl. Phys.*, 12:155–164, 1941.
11. M. A. Biot. Theory of elasticity and consolidation for a porous anisotropic solid. *J. Appl. Phys.*, 25:182–185, 1955.
12. M. A. Biot. Theory of finite deformations of porous solids. *Indiana Univ. Math. J.*, 21:597–620, 1972.
13. O. Coussy. *Mechanics of Porous Continua*. John Wiley & Sons, 1995.
14. R. E. Showalter. Poroelastic filtration coupled to Stokes flow. In O. Imanuvilov, G. Leugering, R. Triggiani, and B. Zhang, editors, *Lecture Notes in Pure and Applied Mathematics, vol. 242*, pages 229–241. Chapman & Hall, Boca Raton, 2005.
15. P. Kuberry and H. Lee. A decoupling algorithm for fluid-structure interaction problems based on optimization. *Comput. Methods. Appl. Mech. Engrg.*, 267: 594–605, 2013.
16. V.J. Ervin, E.W. Jenkins, and H. Lee, Approximation of the Stokes-Darcy system by optimization, J. Sci. Comput., 59;775–794, 2014.
17. Y. Saad, Iterative Methods for Sparse Linear Systems, Second Edition, SIAM, Philadelphia, PA, 2003. MR1990645.

Study of Discrete Scattering Operators for Some Linear Kinetic Models

Yanping Chen, Zheng Chen, Yingda Cheng, Adrianna Gillman
and Fengyan Li

Abstract In this paper, we consider spatially homogeneous linear kinetic models arising from semiconductor device simulations and investigate how various deterministic numerical methods approximate their scattering operators. In particular, methods including first and second order discontinuous Galerkin methods, a first order collocation method, a Fourier-collocation spectral method, and a Nyström method are examined when they are applied to one-dimensional models with singular or continuous scattering kernels. Mathematical properties are discussed for the corresponding discrete scattering operators. We also present numerical experiments to demonstrate the performance of these methods. Understanding how the scattering operators are approximated can provide insights into designing efficient algorithms for simulating kinetic models and for the implicit discretizations of the problems in the presence of multiple scales.

Y. Chen
Department of Mathematical Sciences, The University of Texas at Dallas,
Dallas, TX 75252, USA
e-mail: yxc110030@utdallas.edu

Z. Chen
Computer Science and Mathematics Division, Oak Ridge National Laboratory,
Oak Ridge, TN 37831, USA
e-mail: chenz1@ornl.gov

Y. Cheng
Department of Mathematics, Michigan State University,
East Lansing, MI 48824, USA
e-mail: ycheng@math.msu.edu

A. Gillman
Computational and Applied Mathematics Department, Rice University,
Houston, TX 77005, USA
e-mail: adrianna.gillman@rice.edu

F. Li (✉)
Department of Mathematical Sciences, Rensselaer Polytechnic Institute,
Troy, NY 12180, USA
e-mail: lif@rpi.edu

© Springer Science+Business Media New York 2016
S.C. Brenner (ed.), *Topics in Numerical Partial Differential Equations
and Scientific Computing*, The IMA Volumes in Mathematics
and its Applications 160, DOI 10.1007/978-1-4939-6399-7_5

1 Introduction

Kinetic models arise in many applications such as rarefied gas dynamics, plasma physics, nuclear engineering, semiconductor device design, traffic networking, and swarming. Such models evolve the probability distribution function of one or multiple species of particles, with or without forces from external or self-consistent fields. They can describe mesoscopic phenomena lying in between the microscopic particle dynamics governed by fundamental laws such as the Newton's laws of motion, and macroscopic dynamics described by continuum models. The numerical challenges often come from high dimensionality, various collision (or scattering) operators which can be multifold integrals or singular, and multiple scales in time or phase space.

In this work, we consider some simple one-dimensional linear kinetic models with either singular or continuous scattering operators, and investigate mathematically and/or computationally the properties of several deterministic numerical discretizations. They include first and second order discontinuous Galerkin methods, a first order collocation method, a Fourier-collocation spectral method, and a Nyström method. Lots of efforts have been put in the literature for simulations of semiconductor Boltzmann equations from algorithm and application points of view, for example, computations by spectral methods [6], finite difference methods [2, 4], and discontinuous Galerkin method [3]. In this paper, we are particularly concerned with characterizing and examining how various numerical methods capture the equilibriums. Since only spatially homogeneous models are considered, what we examine here is essentially on how the scattering operators are approximated numerically. Such study is important for understanding numerical approximations for scattering operators which are a key part in any collisional kinetic model and can provide insights into designing efficient algorithms for numerical simulations and also for implicit discretizations of the problems in the presence of multiple scales.

Let us start with the models. Consider a one-dimensional electron-phonon scattering model [5, 8, 10]

$$\frac{\partial f(k,t)}{\partial t} = \hat{S}[f](k,t) = \int_{-\infty}^{\infty} \Big(S(k',k)f(k',t) - S(k,k')f(k,t) \Big) dk', \quad (1)$$

which arises from semiconductor device design. Here, $f(k,t)$ is the probability distribution function of electrons with wave number k at time t, $\hat{S}$ is the scattering operator, and $S(k,k')$ is the scattering kernel which gives the transfer rate of electrons scattering from state k to k'. Note that the space variable x is omitted and the equation (1) is space homogeneous.

The first problem we will focus on is the governing equation (1) that models both the inelastic and elastic scattering, and the scattering kernel is defined as

$$S(k, k') = \sum_{v \in \{-1,0,1\}} s_v(E(k), E(k'))\delta\Big(E(k) - E(k') + v\varepsilon_p\Big). \tag{2}$$

Here $E(k)$ is the energy of the electron with wave number k, $s_v\big(E(k), E(k')\big)$ is the transfer rate from k to k' by absorbing ($v = 1$) or emitting ($v = -1$) a phonon with an energy $\varepsilon_p > 0$, or by keeping the energy unchanged ($v = 0$). And $\delta(\cdot)$ is the Dirac δ function. It is assumed $s_v(\cdot, \cdot) > 0$ with $v = \pm 1$, and $s_0(\cdot, \cdot) \geq 0$. We consider the Kane energy band, with the energy function $E(k)$ satisfying

$$E(k)(1 + \alpha E(k)) = k^2/2 \tag{3}$$

and the nonparabolicity factor $\alpha \geq 0$ is some constant parameter. We also define

$$\mathcal{K}_\alpha(E) = \sqrt{2E(1 + \alpha E)}.$$

The energy function $E(k)$ is nonnegative and it is an even function of k. When $\alpha = 0$, it corresponds to the quadratic energy band.

With $T > 0$ being any given lattice temperature, the following distribution function

$$f^G(k) = \exp\left(-\frac{E(k)}{T}\right) \tag{4}$$

defines an equilibrium of our model, under the assumption

$$s_1(E(k) - \varepsilon_p, E(k)) = s_{-1}(E(k), E(k) - \varepsilon_p)\exp\left(-\frac{\varepsilon_p}{T}\right). \tag{5}$$

This assumption is made throughout this paper, and it ensures the detailed balance principle $S(k', k)f^G(k') = S(k, k')f^G(k)$. For the quadratic energy (3) with $\alpha = 0$, the equilibrium (4) after normalization is a Gaussian distribution. Following a similar analysis as in [9], any equilibrium of our model is given by

$$f^e(k) = f^G(k)h(E(k)), \tag{6}$$

where $h(E)$ is some periodic function of period ε_p. The inclusion of an ε_p-periodic function factor $h(E)$ in an equilibrium is due to the δ-type scattering rule in (2).

The model we have described so far, defined in (1), (2), (3) with the assumption (5), involves a scattering kernel with δ-type singularity. In this work, we will also examine a model which is defined by (1) with a continuous scattering kernel

$$S(k, k') = \sigma(k, k')M(k') \tag{7}$$

where $\sigma(k, k') = \sigma(k', k) \geq 0$ and $M(k) = \frac{1}{\sqrt{2\pi}} \exp(-\frac{k^2}{2T})$. For any given temperature $T > 0$, this model has a unique Gaussian-type equilibrium $M(k)$ (up to a constant factor).

In [7], a first order finite volume method was introduced for the linear kinetic model (1) with (2), (3) and the assumption (5), when the energy band is quadratic ($\alpha = 0$) without the elastic collision ($s_0 = 0$). A detailed study of the scattering matrix which approximates the scattering operator was performed. In particular, the eigenvalues of the scattering matrix were proven to be nonpositive, showing the stability of the numerical scheme. The dependence of the geometric multiplicity of the zero eigenvalue on the choice of the mesh grids was established based on linear algebra tools. Such theory was extended in [11] to more general models, such as those with general energy band (including Kane energy), with anisotropic scattering, and in higher dimensions. Our aim in this paper is to perform a thorough numerical study of the model with either a singular or continuous scattering kernel by considering more general methods, including higher order Galerkin-type method, collocation methods of low or high order accuracy. We are particularly concerned with the scattering matrix resulted from different types of discretization, and the interpretation of the numerical results when compared with their continuous counterparts in the models.

The rest of the paper is organized as follows. In Section 2, the kinetic model (1) with a singular scattering kernel (2)-(3) is considered. More specifically, a first order finite volume method as in [7, 11], which is also a first order discontinuous Galerkin (DG) method, is formulated in Section 2.2.1. Mathematical properties of the numerical scheme as well as the scattering matrix are reviewed, followed with some discussions. More general numerical methods, including a second order DG method, a first order collocation method, and a Fourier-collocation spectral method are formulated in Sections 2.2.2, 2.2.3 and 2.2.4, respectively. It turns out it is nontrivial to extend the algebraic analysis in [7] to more general numerical discretizations. Instead, we rely on extensive numerical experiments to understand these methods, see Section 2.3. In Section 3, the kinetic model (1) with a continuous scattering kernel (7) is considered, for which a Nyström discretization is introduced and tested numerically. A detailed summary and concluding remarks are made in Section 4.

2 Numerical Methods for Singular Scattering Kernels

In this section, the kinetic model (1) will be considered with a singular scattering kernel (2) and (3). We will start with rewriting the equation. We then formulate a first order discontinuous Galerkin (DG) method, which is also a first order finite volume method, a second order DG method, a first order collocation method, and a Fourier-collocation spectral method. For the first order DG method, we will also discuss the mathematical properties of the discrete scattering operator.

2.1 Reformulation of the Model

Before introducing numerical methods, we first reformulate the scattering terms in our model to formally remove the δ-type singularity. Details will be given only for one term associated with the inelastic scattering, and the remaining terms can be treated similarly. Recall the definition of the composition of the δ-function with a differentiable function $z(\cdot)$,

$$\int_{-\infty}^{\infty} \delta(z(x))v(x)dx = \sum_{x_\star \in \{y:z(y)=0\}} \frac{v(x_\star)}{|z'(x_\star)|},$$

and using this, one gets

$$\begin{aligned}
\mathscr{R}_1[f](k,t) &= \int_{-\infty}^{\infty} s_1(E(k'), E(k))\delta(E(k') - E(k) + \varepsilon_p)f(k',t)dk' \\
&= \sum_{k_\star \in \{k_\star : E(k_\star) = E(k) - \varepsilon_p\}} \frac{s_1(E(k_\star), E(k))f(k_\star, t)}{|E'(k_\star)|}.
\end{aligned} \tag{8}$$

Notation wise, one should understand that for any k with $E(k) - \varepsilon_p < 0$, the corresponding term in (8) is excluded.

With $E(k_\star) = E(k) - \varepsilon_p$, one can easily verify

$$k_\star = \pm\sqrt{2E(k_\star)(1 + \alpha E(k_\star))} = \pm\mathscr{K}_\alpha(E(k) - \varepsilon_p), \tag{9}$$

$$E'(k_\star) = \frac{k_\star}{1 + 2\alpha E(k_\star)} = \frac{\pm\mathscr{K}_\alpha(E(k) - \varepsilon_p)}{1 + 2\alpha(E(k) - \varepsilon_p)}. \tag{10}$$

Combining (8)–(10), we have

$$\mathscr{R}_1[f](k,t) = \frac{s_1(E, E + \varepsilon_p)}{\mathscr{K}_\alpha(E)/(1 + 2\alpha E)}\Big(f(\mathscr{K}_\alpha(E), t) + f(-\mathscr{K}_\alpha(E), t)\Big)\Big|_{E=E(k)-\varepsilon_p}. \tag{11}$$

Following the similar derivation for other terms, our model (1)–(3) with the singular scattering kernel is reformulated as below,

$$\frac{\partial f(k,t)}{\partial t} = \hat{S}[f](k,t) = \sum_{m=1}^{4} \mathscr{R}_m[f](k,t), \tag{12}$$

where

$$\mathscr{R}_2[f](k,t) = \frac{s_{-1}(E, E - \varepsilon_p)}{\mathscr{K}_\alpha(E)/(1 + 2\alpha E)}\Big(f(\mathscr{K}_\alpha(E), t) + f(-\mathscr{K}_\alpha(E), t)\Big)|_{E = E(k) + \varepsilon_p}$$

$$\mathscr{R}_3[f](k,t) = \frac{s_0(E, E)}{\mathscr{K}_\alpha(E)/(1 + 2\alpha E)}\Big(f(\mathscr{K}_\alpha(E), t) + f(-\mathscr{K}_\alpha(E), t)\Big)|_{E = E(k)}$$

$$\mathscr{R}_4[f](k,t) = -2f(k,t)\left(\frac{s_1(E - \varepsilon_p, E)}{\mathscr{K}_\alpha(E)/(1 + 2\alpha E)}|_{E = E(k) + \varepsilon_p}\right.$$

$$+ \frac{s_{-1}(E + \varepsilon_p, E)}{\mathscr{K}_\alpha(E)/(1 + 2\alpha E)}|_{E = E(k) - \varepsilon_p}$$

$$\left.+ \frac{s_0(E, E)}{\mathscr{K}_\alpha(E)/(1 + 2\alpha E)}|_{E = E(k)}\right).$$

2.2 Numerical Methods

2.2.1 First Order Discontinuous Galerkin Method

In this subsection, we will describe a discontinuous Galerkin (DG) method using piecewise constant discrete space to numerically approximate the reformulated model (12). The method is also the first order finite volume method studied in [7, 11] in the absence of the elastic scattering term, namely when $s_0 = 0$.

We start with introducing some notation. Let $[-K_{\max}, K_{\max}]$ be the computational domain, with the assumption that the exact solution is zero in the machine accuracy level outside this domain. Let $0 = k_{1/2} < k_{3/2} < \cdots < k_{N+1/2} = K_{\max}$ be a partition of $[0, K_{\max}]$, and define $I_i = [k_{i-1/2}, k_{i+1/2}]$, $\Delta k_i = k_{i+1/2} - k_{i-1/2}$, $\forall i \in \mathscr{N}^+ = \{1, 2, \cdots, N\}$, and $\Delta k = \max_{1 \le i \le N} \Delta k_i$. For the left-half domain $[-K_{\max}, 0]$, a "symmetric" mesh is introduced with $I_{-i} = [k_{-i-1/2}, k_{-i+1/2}]$, and $k_{-i-1/2} = -k_{i+1/2}$, $i \in \mathscr{N}^+$. In terms of the energy variable, we define $E_{\max} = E(K_{\max})$, $E_{i-1/2} = E(k_{i-1/2})$, $i = 1, \cdots, N + 1$, $\Omega_i = [E_{i-1/2}, E_{i+1/2}]$, $\Delta E_i = E_{i+1/2} - E_{i-1/2}$, $i \in \mathscr{N}^+$, and $\Delta E = \max_{1 \le i \le N} \Delta E_i$. We also use $\Omega_i \pm \varepsilon_p = \{E \pm \varepsilon_p : E \in \Omega_i\}$ and $\mathscr{N} = \{-N, \cdots, -2, -1, 1, 2, \cdots, N\}$.

To formulate the method, we approximate $f(k, t)$ by a piecewise constant function $f_h(k, t)$, namely $f_h(\cdot, t) \in V_h = V_h^0 = \{g : g|_{I_i} \in P^0(I_i), \forall i \in \mathscr{N}\}$, satisfying

$$\int_{I_i} \frac{\partial f_h(k, t)}{\partial t}\phi(k)dk = \int_{I_i} \hat{S}[f_h](k, t)\phi(k)dk = \sum_{m=1}^{4} \int_{I_i} \mathscr{R}_m[f_h](k, t)\phi(k)dk \quad (13)$$

for any $\phi \in V_h$ and any $i \in \mathscr{N}$. Here and below $P^r(I_i)$ is the set of polynomials on I_i of degree r. This scheme, in its finite volume form, is also given as (14) with $\phi = 1$,

$$\int_{I_i} \frac{\partial f_h(k,t)}{\partial t} dk = \int_{I_i} \hat{S}[f_h](k,t)dk = \sum_{m=1}^{4} \int_{I_i} \mathscr{R}_m[f_h](k,t)dk. \qquad (14)$$

1.) The scheme in its algebraic form

Next, we will convert the scheme (14) into its algebraic form. To do so, we represent the numerical solution as $f_h(k,t)|_{I_i} = f_i(t)$, with $f_i(t) = \frac{1}{\Delta k_i} \int_{I_i} f_h(k,t)dk$ which approximates the cell average of the exact solution $f(k,t)$ over I_i, $\forall i \in \mathcal{N}$. It is straightforward to get,

$$\int_{I_i} \frac{\partial f_h(k,t)}{\partial t} dk = \frac{d}{dt}(\Delta k_i f_i(t)). \qquad (15)$$

To proceed with the remaining terms related to the scattering operator, we will take a change of variable from k to E. With the relation between the velocity k and the energy E in (3), we have $dk = (1 + 2\alpha E)/\mathscr{K}_\alpha(E)\, dE$ and

$$\int_{I_i} z(E(k))dk = \int_{\Omega_{|i|}} z(E)\frac{1 + 2\alpha E}{\mathscr{K}_\alpha(E)} dE$$

for any given function $z(\cdot)$.

For the first term on the right-hand side of (14), we have

$$\int_{I_i} \mathscr{R}_1[f_h](k,t)dk = \int_{\Omega_{|i|}} \mathscr{R}_1[f_h](\mathscr{K}_\alpha(E),t)\frac{1 + 2\alpha E}{\mathscr{K}_\alpha(E)} dE$$

$$= \int_{\Omega_{|i|-\varepsilon_p}} s_1(E, E+\varepsilon_p)\frac{1 + 2\alpha E}{\mathscr{K}_\alpha(E)} \cdot \frac{1 + 2\alpha(E+\varepsilon_p)}{\mathscr{K}_\alpha(E+\varepsilon_p)} \Big(f_h(\mathscr{K}_\alpha(E),t) + f_h(-\mathscr{K}_\alpha(E),t)\Big)dE$$

$$= \int_{\Omega_{|i|-\varepsilon_p}} s_1(E, E+\varepsilon_p)\frac{1 + 2\alpha E}{\mathscr{K}_\alpha(E)} \cdot \frac{1 + 2\alpha(E+\varepsilon_p)}{\mathscr{K}_\alpha(E+\varepsilon_p)} \sum_{j\in\mathcal{N}} \chi_{\Omega_{|j|}}(E)f_j(t)dE$$

$$= \sum_{j\in\mathcal{N}} f_j(t)r_{i,j}^{(1)}, \qquad (16)$$

with

$$r_{i,j}^{(1)} = \int_{(\Omega_{|i|-\varepsilon_p})\cap\Omega_{|j|}} s_1(E, E+\varepsilon_p)\frac{1 + 2\alpha E}{\mathscr{K}_\alpha(E)} \cdot \frac{1 + 2\alpha(E+\varepsilon_p)}{\mathscr{K}_\alpha(E+\varepsilon_p)} dE. \qquad (17)$$

Following similar derivation, we can further get $\int_{I_i} \mathscr{R}_m[f_h](k,t)dk = \sum_{j\in\mathcal{N}} f_j(t)r_{i,j}^{(m)}$, $m = 2, 3$ with

$$r_{i,j}^{(2)} = \int_{(\Omega_{|i|+\varepsilon_p})\cap\Omega_{|j|}} s_{-1}(E, E-\varepsilon_p)\frac{1 + 2\alpha E}{\mathscr{K}_\alpha(E)} \cdot \frac{1 + 2\alpha(E-\varepsilon_p)}{\mathscr{K}_\alpha(E-\varepsilon_p)} dE$$

$$r_{i,j}^{(3)} = \int_{\Omega_{|i|}\cap\Omega_{|j|}} s_0(E, E)\left(\frac{1 + 2\alpha E}{\mathscr{K}_\alpha(E)}\right)^2 dE = \delta_{|i|,|j|}\int_{\Omega_{|i|}} s_0(E, E)\left(\frac{1 + 2\alpha E}{\mathscr{K}_\alpha(E)}\right)^2 dE,$$

and $\int_{I_i} \mathscr{R}_4[f_h](k, t)dk = -2f_i(t)\hat{\lambda}_i$, with

$$
\hat{\lambda}_i = \int_{\Omega_{|i|}} \left(\frac{s_1(E, E + \varepsilon_p)(1 + 2\alpha(E + \varepsilon_p))}{\mathscr{K}_\alpha(E + \varepsilon_p)} + \frac{s_{-1}(E, E - \varepsilon_p)(1 + 2\alpha(E - \varepsilon_p))}{\mathscr{K}_\alpha(E - \varepsilon_p)} \right)
$$
$$
\frac{1 + 2\alpha E}{\mathscr{K}_\alpha(E)} dE + \int_{\Omega_{|i|}} s_0(E, E) \left(\frac{1 + 2\alpha E}{\mathscr{K}_\alpha(E)} \right)^2 dE. \tag{18}
$$

Here δ_{ij} is the Kronecker-δ function.

By combining what we have so far in (14)–(18), the proposed first order DG scheme for the model (12) with the singular scattering kernel is converted to its algebraic form,

$$
\frac{d}{dt}(\Delta k_i f_i) = -2\lambda_i(\Delta k_i f_i) + \sum_{j \in \mathcal{N}} s_{i,j}(\Delta k_j f_j), \quad \forall i \in \mathcal{N}, \tag{19}
$$

where

$$
s_{i,j} = \frac{1}{\Delta k_j}(r_{i,j}^{(1)} + r_{i,j}^{(2)} + r_{i,j}^{(3)}) \quad \text{and} \quad \lambda_i = \frac{1}{\Delta k_i}\hat{\lambda}_i. \tag{20}
$$

2.) Properties of the scheme

With $s_v(\cdot, \cdot) > 0$, $v = \pm 1$ and $s_0(\cdot, \cdot) \geq 0$, one can easily see that all the coefficients in the linear algebraic system (19) are nonnegative, more specifically,

$$
s_{i,j} \geq 0, \quad \lambda_i > 0, \quad \forall i, j \in \mathcal{N}. \tag{21}
$$

They also have some symmetry property, namely

$$
\lambda_i = \lambda_{-i}, \quad s_{i,j} = s_{-i,j} = s_{i,-j} = s_{-i,-j}, \quad \forall i, j \in \mathcal{N} \tag{22}
$$

due to that the energy $E(k)$ is an even function in k and the mesh is "symmetrically" defined.

Now we introduce

$$
\Lambda = \text{diag}\{\lambda_1, \cdots, \lambda_N\}, \quad S = (s_{i,j})_{i,j \in \mathcal{N}^+}, \tag{23}
$$

and $\mathbf{f}_- = [\Delta k_1 f_{-1}, \cdots, \Delta k_N f_{-N}]^T$, $\mathbf{f}_+ = [\Delta k_1 f_1, \cdots, \Delta k_N f_N]^T$; then, the proposed scheme in (19) can be written as

$$
\frac{d}{dt}\begin{bmatrix} \mathbf{f}_- \\ \mathbf{f}_+ \end{bmatrix} = \mathbb{S}\begin{bmatrix} \mathbf{f}_- \\ \mathbf{f}_+ \end{bmatrix} = \left(-2\begin{bmatrix} \Lambda & 0 \\ 0 & \Lambda \end{bmatrix} + \begin{bmatrix} S & S \\ S & S \end{bmatrix} \right)\begin{bmatrix} \mathbf{f}_- \\ \mathbf{f}_+ \end{bmatrix}. \tag{24}
$$

The matrix $\mathbb{S}$ is the discrete matrix corresponding to the discrete scattering operator. If we further define $\mathbf{g} = (\mathbf{f}_- + \mathbf{f}_+)/2 \in \mathbb{R}^N$, $\mathbf{h} = (\mathbf{f}_+ - \mathbf{f}_-)/2 \in \mathbb{R}^N$, and $M = 2(S - \Lambda)$, the linear system (24) can be decoupled into two systems of halved size,

$$\frac{d}{dt}\mathbf{g} = M\mathbf{g}, \qquad \frac{d}{dt}\mathbf{h} = -2\Lambda\mathbf{h}. \tag{25}$$

It is easy to see solving the proposed scheme (19) (or (24)) is equivalent to solving (25). In next lemma, we will summarize more properties of $\mathbb{S}$, M and Λ.

Lemma 2.1 *1.) Λ is nonsingular.*
2.) The eigenvalues of $\mathbb{S}$ consist of all eigenvalues of M and of -2Λ. Hence, the dimensions of $ker(\mathbb{S})$ and $ker(M)$ are the same.
3.) $\mathbf{g} \in ker(M) \Leftrightarrow [\mathbf{g}^\top, \mathbf{g}^\top]^\top \in ker(\mathbb{S})$.

The proof is straightforward, and it is omitted here. Based on the properties in this lemma, we can see that to address the types of questions as in [7] for the scattering matrix $\mathbb{S}$, such as the dimension of the null space of $\mathbb{S}$, the sign of the real part of the eigenvalues of $\mathbb{S}$, it is equivalent to ask similar questions to the reduced scattering matrix M. On the other hand, to get numerical solution $f_h(k, t)$ at any time t, one would have to work with both equations in (25) or with equation (24).

Next we will verify directly that the scheme given above has mass conservation property. An important consequence is that the column sum of M is zero. This property ensures zero is an eigenvalue of M, and it was also extensively used in analyzing M in [7]. Such property is usually not possessed by collocation-type methods. Instead with collocation methods, zero eigenvalue of the scattering operator can be approximated by nonzero numerical eigenvalues (see numerical results in Section 2.3).

Lemma 2.2 *Suppose the numerical solution $f_h(k, t)$ has compact support in $[-K_{max}, K_{max}]$, then the proposed scheme (14) satisfies mass conservation, namely*

$$\frac{d}{dt}\int_{-K_{max}}^{K_{max}} f_h(k, t)dk = \frac{d}{dt}\sum_{i \in \mathcal{N}} \Delta k_i f_i(t) = 0. \tag{26}$$

Moreover $\sum_{i \in \mathcal{N}^+} M_{ij} = 0$, $\forall j \in \mathcal{N}^+$.

Proof Based on the formulas for $s_{i,j}$ and λ_j in (20) as well as the symmetry relation in (22), one can verify

$$\frac{1}{2}\sum_{i \in \mathcal{N}} s_{i,j} = \sum_{i \in \mathcal{N}^+} s_{i,j} = \lambda_j, \quad \forall j \in \mathcal{N}^+ \tag{27}$$

and hence $\sum_{i \in \mathcal{N}^+} M_{ij} = 2\left(\sum_{i \in \mathcal{N}^+} s_{i,j} - \lambda_j\right) = 0$ for all $\forall j \in \mathcal{N}^+$.

With this and (19), we have

$$\frac{d}{dt}\int_{-K_{max}}^{K_{max}} f_h(k,t)dk = \frac{d}{dt}\sum_{i\in\mathcal{N}}\Delta k_i f_i(t) = -2\sum_{i\in\mathcal{N}}\lambda_i(\Delta k_i f_i) + \sum_{i\in\mathcal{N}}\sum_{j\in\mathcal{N}} s_{i,j}(\Delta k_j f_j)$$

$$= -2\sum_{i\in\mathcal{N}}\lambda_i(\Delta k_i f_i) + \sum_{j\in\mathcal{N}}(\sum_{i\in\mathcal{N}} s_{i,j})(\Delta k_j f_j)$$

$$= -2\sum_{i\in\mathcal{N}}\lambda_i(\Delta k_i f_i) + \sum_{j\in\mathcal{N}} 2\lambda_j(\Delta k_j f_j) = 0. \tag{28}$$

$\square$

In next theorem, we summarize the main results which were proved for the (reduced) scattering matrix M in [7, 11] when $s_0 = 0$. s_0 being nonzero does not pose new difficulty.

Theorem 2.3 *1.)* $M_{ij} \geq 0$ *for* $i \neq j$, $M_{ii} < 0$, *and* $M^\top$ *is weakly diagonally dominant. Each nonzero eigenvalue of M has a negative real part.*

2.) $M_{ij} > 0 \Leftrightarrow M_{ji} > 0$. *In addition, there exists a unique positive integer s and a permutation matrix P such that*

$$M = P^\top \begin{bmatrix} M_1 & & \\ & \ddots & \\ & & M_s \end{bmatrix} P, \tag{29}$$

where each $M_i \in \mathbb{R}^{r_i \times r_i}$, $i = 1, \cdots, s$ is irreducible. Moreover $rank(M_i) = r_i - 1$, and this implies $rank(M) = N - s$ and $dim(ker(M)) = s$. Let $\mathbf{g}_i \in null(M_i)$ be nonzero for any i, all entries of $\mathbf{g}_i$ have the same sign.

3.) The fact that $dim(ker(M)) = s$ can be equivalently characterized by the following property of the mesh: there exists $E_1^\star, \cdots E_s^\star \in [0, \varepsilon_p)$ with $E_1 < \cdots < E_s^\star$, such that

$$\{E_i^\star + \tau\varepsilon_p : E_i^\star + \tau\varepsilon_p \leq E_{max}, \tau \in \mathbb{N}\} \subseteq \{E_{j-1/2} : j = 1, 2, \cdots N+1\}, \quad i = 1, \cdots, s. \tag{30}$$

From the theorem, one can see that there is no nonzero eigenvalue of M with real positive part, and this implies the stability of the scheme and ensures the correct decay behavior of the numerical solution over long time period. One can always find a set of basis for the null space of M such that each basis vector in nonnegative. In addition, the geometric multiplicity of zero eigenvalue being s, hence $dim(ker(M)) = s$, can be fully characterized by the choice of the mesh grids. To further understand the mesh condition in (30), recall our model admits infinitely many equilibriums (6), and the presence of an ε_p-periodic function factor $h(E(k))$ is due to the δ-type scattering rule in the model. With this, the behaviors of an equilibrium $f^e(k)$ at k and k' are related only when $E(k) = E(k') + v\varepsilon_p$, with $v = -1, 0, 1$. The statement in 3.) implies that the dimension s of the null space of M is same as the total number of decoupled subregions of the energy domain under the scattering rule on the *numerical* level. (This is best illustrated by Figure 1 in [7].) Such result is not hard to get intuitively, and it is mathematically justified by the Theorem above for the first order DG method. It turns out similar analysis is nontrivial to establish for other numerical discretizations considered in Section 2.2. Without any analysis available, in order to understand how the scattering rule determined by each numerical discretization of the model decouples the energy domain, to what extent the numerical discretization captures the equilibriums of the scattering operator, we will numerically examine the null space of M or the steady state of the discretized system, see Section 2.3.

Remark 2.4 In practice, uniform meshes in the energy variable E are often used with $\Delta E_i = \Delta E, \forall i \in \mathcal{N}^+$. In such situation, if $\varepsilon_p/\Delta E = n \in \mathbb{Z}^+$, we have $dim(ker(M)) = n$; if $\varepsilon_p/\Delta E$ is not an integer, then $dim(ker(M)) = 1$.

Remark 2.5 The mass conservation property is one of the keys for the results in the above theorem. It is ensured by the relation (27). To implement the proposed scheme, if $s_{i,j}$ and $\lambda_i, \forall i, j \in \mathcal{N}^+$ are computed independently using numerical quadrature, this relation will hold only up to the accuracy of the quadrature formulas. In our actual implementation, $\{s_{i,j}\}_{i,j \in \mathcal{N}^+}$ are computed first, then λ_j is obtained based on (27); hence, the mass conservation is enforced.

2.2.2 Second Order Discontinuous Galerkin Method

Following the same notation for the computational domain and the mesh as in Section 2.2.1, we introduce the discrete space

$$V_h = V_h^1 = \{g : g|_{I_i} \in P^1(I_i), \forall i \in \mathcal{N}\} \tag{31}$$

which consists of piecewise linear polynomials with respect to the mesh. We then approximate the solution $f(k, t)$ by $f_h(\cdot, t) \in V_h$, satisfying

$$\int_{I_i} \frac{\partial f_h(k, t)}{\partial t} \phi(k) dk = \int_{I_i} \hat{S}[f_h](k, t)\phi(k) dk = \sum_{m=1}^{4} \int_{I_i} \mathcal{R}_m[f_h](k, t)\phi(k) dk \tag{32}$$

for any $\phi \in V_h$ and $i \in \mathcal{N}$. This results in a (formally) second order DG method.

To convert our scheme into its algebraic form, suppose $\phi_i^0(k)$ and $\phi_i^1(k)$ are the basis functions of $P^1(I_i)$, and the numerical solution is represented as $f_h(k, t)|_{I_i} = f_i^0(t)\phi_i^0(k) + f_i^1(t)\phi_i^1(k)$, with $f_i^0(t)$ and $f_i^1(t)$ to be determined by the scheme (32). With the test function $\phi \in V_h$ in (32) taken to be $\phi|_{I_i} = g_i^0\phi_i^0(k) + g_i^1\phi_i^1(k)$, the term on the left-hand side becomes

$$\int_{I_i} \frac{\partial f_h(k, t)}{\partial t}\phi(k) dk = [g_i^0, g_i^1]A_i \frac{d}{dt}\begin{bmatrix} f_i^0 \\ f_i^1 \end{bmatrix},$$

with

$$A_i = \int_{I_i} \begin{bmatrix} (\phi_i^0)^2 & \phi_i^1 \phi_i^0 \\ \phi_i^0 \phi_i^1 & (\phi_i^1)^2 \end{bmatrix} dk.$$

For the first term on the right-hand side of (32), we have

$$\int_{I_i} \mathscr{R}_1[f_h](k,t)\phi(k)dk = \int_{\Omega_{|i|}} \mathscr{R}_1[f_h](\mathscr{K}_\alpha(E),t)\frac{1+2\alpha E}{\mathscr{K}_\alpha(E)}\phi\Big(\text{sign}(i)\,\mathscr{K}_\alpha(E)\Big)dE$$

$$= \int_{\Omega_{|i|}-\varepsilon_p} \mathscr{R}_1[f_h](\mathscr{K}_\alpha(E+\varepsilon_p),t)\frac{1+2\alpha(E+\varepsilon_p)}{\mathscr{K}_\alpha(E+\varepsilon_p)}\phi\Big(\text{sign}(i)\,\mathscr{K}_\alpha(E+\varepsilon_p)\Big)dE$$

$$= \int_{\Omega_{|i|}-\varepsilon_p} s_1(E,E+\varepsilon_p)\frac{1+2\alpha E}{\mathscr{K}_\alpha(E)}\cdot\frac{1+2\alpha(E+\varepsilon_p)}{\mathscr{K}_\alpha(E+\varepsilon_p)}\Big(f_h(\mathscr{K}_\alpha(E),t)$$

$$+ f_h(-\mathscr{K}_\alpha(E),t)\Big)\phi\Big(\text{sign}(i)\,\mathscr{K}_\alpha(E+\varepsilon_p)\Big)dE. \tag{33}$$

Note that

$$f_h(\mathscr{K}_\alpha(E),t) + f_h(-\mathscr{K}_\alpha(E),t)$$
$$= \sum_{j\in\mathscr{N}} \chi_{\Omega_{|j|}}(E)\left(f_j^0\phi_j^0(k) + f_j^1\phi_j^1(k)\right)|_{k=\text{sign}(j)\,\mathscr{K}_\alpha(E)}, \tag{34}$$

then

$$\int_{I_i} \mathscr{R}_1[f_h](k,t)\phi(k)dk = \sum_{j\in\mathscr{N}} [g_i^0, g_i^1]S_{i,j}^1 \begin{bmatrix} f_j^0 \\ f_j^1 \end{bmatrix}, \tag{35}$$

with

$$S_{i,j}^1 = \int_{(\Omega_{|i|}-\varepsilon_p)\cap\Omega_{|j|}} s_1(E,E+\varepsilon_p)\frac{(1+2\alpha E)}{\mathscr{K}_\alpha(E)}\cdot\frac{(1+2\alpha(E+\varepsilon_p))}{\mathscr{K}_\alpha(E+\varepsilon_p)}\begin{bmatrix} \phi_j^0(\Delta)\phi_i^0(\Delta_1) & \phi_j^1(\Delta)\phi_i^0(\Delta_1) \\ \phi_j^0(\Delta)\phi_i^1(\Delta_1) & \phi_j^1(\Delta)\phi_i^1(\Delta_1) \end{bmatrix}dE$$

and $\Delta = \text{sign}(j)\,\mathscr{K}_\alpha(E)$, $\Delta_1 = \text{sign}(i)\,\mathscr{K}_\alpha(E+\varepsilon_p)$. Similarly,

$$\int_{I_i} \mathscr{R}_m[f_h](k,t)\phi(k)dk = \sum_{j\in\mathscr{N}} [g_i^0, g_i^1]S_{i,j}^m \begin{bmatrix} f_j^0 \\ f_j^1 \end{bmatrix} \tag{36}$$

for $m = 2, 3$, with

$$S_{i,j}^2 = \int_{(\Omega_{|i|}+\varepsilon_p)\cap\Omega_{|j|}} s_{-1}(E,E-\varepsilon_p)\frac{(1+2\alpha E)}{\mathscr{K}_\alpha(E)}\cdot\frac{(1+2\alpha(E-\varepsilon_p))}{\mathscr{K}_\alpha(E-\varepsilon_p)}\begin{bmatrix} \phi_j^0(\Delta)\phi_i^0(\Delta_2) & \phi_j^1(\Delta)\phi_i^0(\Delta_2) \\ \phi_j^0(\Delta)\phi_i^1(\Delta_2) & \phi_j^1(\Delta)\phi_i^1(\Delta_2) \end{bmatrix}dE,$$

$$S_{i,j}^3 = \delta_{|i|,|j|}\int_{\Omega_{|i|}} s_0(E,E)\Big(\frac{1+2\alpha E}{\mathscr{K}_\alpha(E)}\Big)^2 \begin{bmatrix} \phi_j^0(\Delta)\phi_i^0(\Delta_3) & \phi_j^1(\Delta)\phi_i^0(\Delta_3) \\ \phi_j^0(\Delta)\phi_i^1(\Delta_3) & \phi_j^1(\Delta)\phi_i^1(\Delta_3) \end{bmatrix}dE,$$

and $\Delta_2 = \text{sign}(i) \, \mathcal{K}_\alpha(E - \varepsilon_p)$, $\Delta_3 = \text{sign}(i) \, \mathcal{K}_\alpha(E)$. Moreover,

$$\int_{I_i} \mathcal{R}_4[f_h](k,t)\phi(k)dk = -2[g_i^0, g_i^1]\Lambda_i \begin{bmatrix} f_i^0 \\ f_i^1 \end{bmatrix}, \tag{37}$$

with

$$\Lambda_i = \int_{\Omega_{|i|}} \Theta(E) \begin{bmatrix} (\phi_i^0(\Delta_3))^2 & \phi_i^1(\Delta_3)\phi_i^0(\Delta_3) \\ \phi_i^0(\Delta_3)\phi_i^1(\Delta_3) & (\phi_i^1(\Delta_3))^2 \end{bmatrix} dE.$$

Here, $\Theta(E) = \left(\frac{s_1(E, E+\varepsilon_p)(1+2\alpha(E+\varepsilon_p))}{\mathcal{K}_\alpha(E+\varepsilon_p)} + \frac{s_{-1}(E, E-\varepsilon_p)(1+2\alpha(E-\varepsilon_p))}{\mathcal{K}_\alpha(E-\varepsilon_p)} \right) \frac{1+2\alpha E}{\mathcal{K}_\alpha(E)} + s_0(E, E)\left(\frac{1+2\alpha E}{\mathcal{K}_\alpha(E)}\right)^2.$

Now with $S_{i,j} = S_{i,j}^1 + S_{i,j}^2 + S_{i,j}^3$, and the test function ϕ being arbitrary, the scheme becomes

$$A_i \frac{d}{dt} \begin{bmatrix} f_i^0 \\ f_i^1 \end{bmatrix} = -2\Lambda_i \begin{bmatrix} f_i^0 \\ f_i^1 \end{bmatrix} + \sum_{j \in \mathcal{N}} S_{i,j} \begin{bmatrix} f_j^0 \\ f_j^1 \end{bmatrix}, \qquad i \in \mathcal{N}. \tag{38}$$

Next we specify the local basis functions $\{\phi_i^r\}_{i \in \mathcal{N}, r = 1,2}$ as Lagrangian basis, given as

$$\phi_i^0(k) = \frac{1}{\Delta k_i}(k_{i+\frac{1}{2}} - k), \quad \phi_i^1(k) = \frac{1}{\Delta k_i}(k - k_{i-\frac{1}{2}}), \quad \text{if } i > 0, \tag{39}$$

$$\phi_i^0(k) = \frac{1}{\Delta k_{|i|}}(k - k_{i-\frac{1}{2}}), \quad \phi_i^1(k) = \frac{1}{\Delta k_{|i|}}(k_{i+\frac{1}{2}} - k), \quad \text{if } i < 0. \tag{40}$$

With such choice, the local basis functions have certain symmetry,

$$\phi_i^r(k) = \phi_{-i}^r(-k), \qquad r = 0, 1, \quad i \in \mathcal{N}, \tag{41}$$

and so are the elementwise matrices

$$S_{i,j} = S_{i,-j} = S_{-i,j} = S_{-i,-j}, \qquad \Lambda_i = \Lambda_{-i}, \qquad A_i = A_{-i}. \tag{42}$$

If we introduce $\mathbf{f}_- = [f_{-1}^0, f_{-1}^1, \cdots, f_{-N}^0, f_{-N}^1]^T$, $\mathbf{f}_+ = [f_1^0, f_1^1, \cdots, f_N^0, f_N^1]^T$, the scheme (38) can be written more compactly,

$$\begin{bmatrix} A & 0 \\ 0 & A \end{bmatrix} \frac{d}{dt} \begin{bmatrix} \mathbf{f}_- \\ \mathbf{f}_+ \end{bmatrix} = \mathbb{S} \begin{bmatrix} \mathbf{f}_- \\ \mathbf{f}_+ \end{bmatrix} = \left(-2 \begin{bmatrix} \Lambda & 0 \\ 0 & \Lambda \end{bmatrix} + \begin{bmatrix} S & S \\ S & S \end{bmatrix} \right) \begin{bmatrix} \mathbf{f}_- \\ \mathbf{f}_+ \end{bmatrix}. \tag{43}$$

The matrix $A \in \mathbb{R}^{2N \times 2N}$ (resp. $\Lambda \in \mathbb{R}^{2N \times 2N}$) is a $N \times N$ block-diagonal matrix, with its (i,i)th block being A_i (resp. Λ_i). The matrix $S \in \mathbb{R}^{2N \times 2N}$ is a $N \times N$ block-structured matrix, with its (i,j)th block being $S_{i,j}$. And the scheme (43) can be further decoupled into two systems of halved size,

$$A\frac{d}{dt}\mathbf{g} = M\mathbf{g}, \qquad A\frac{d}{dt}\mathbf{h} = -2\Lambda\mathbf{h}. \tag{44}$$

Here $\mathbf{g} = (\mathbf{f}_- + \mathbf{f}_+)/2$, $\mathbf{h} = (\mathbf{f}_+ - \mathbf{f}_-)/2$, and $M = 2(S - \Lambda)$. We can verify directly from the definition that both Λ and the mass matrix A are invertible.

Similar as for the first order DG method, if we are only concerned with the discrete equilibrium such as the dimension of the null space of the scattering matrix $\mathbb{S}$ in (43), it is sufficient to simply consider $A\frac{d}{dt}\mathbf{g} = M\mathbf{g}$ for the same question. For the time-evolving numerical solution $f_h(k, t)$, one needs to work with (43), or equivalently the two equations in (44). On the other hand, it is nontrivial to extend most of the algebraic analysis in [7, 11] to this second order method, for which the involved matrices are of block structure.

What we do know is that the column sum of M is zero, and this again is closely related to the mass conservation of the method, as stated in next lemma.

Lemma 2.6 *Suppose the numerical solution $f_h(k, t)$ has compact support in $[-K_{max}, K_{max}]$, then the proposed scheme (32) satisfies mass conservation, namely*

$$\frac{d}{dt}\int_{-K_{max}}^{K_{max}} f_h(k, t)dk = 0. \tag{45}$$

In addition, the sum of each column of M is zero.

Proof Based on the formulas for Λ_i and $S_{i,j}$, the symmetry in (42), as well as the equality $\phi_i^0 + \phi_i^1 = 1$ on I_i, one can verify

$$[1, 1]\left(-2\Lambda_i + \sum_{j\in\mathcal{N}} S_{j,i}\right) = 0 \tag{46}$$

and the sum of M being 0.

Using (46) as well as (32) with $\phi(k) \equiv 1$, we have

$$\frac{d}{dt}\int_{-K_{max}}^{K_{max}} f_h(k, t)dk = \sum_{i\in\mathcal{N}}\int_{I_i}\frac{\partial f_h(k, t)}{\partial t}dk = \sum_{i\in\mathcal{N}}\int_{I_i}\hat{S}[f_h](k, t)dk$$

$$= \sum_{i\in\mathcal{N}}[1, 1]\left(-2\Lambda_i\begin{bmatrix}f_i^0\\f_i^1\end{bmatrix} + \sum_{j\in\mathcal{N}}S_{j,i}\begin{bmatrix}f_j^0\\f_j^1\end{bmatrix}\right)$$

$$= \sum_{i\in\mathcal{N}}[1, 1]\left(-2\Lambda_i + \sum_{j\in\mathcal{N}}S_{j,i}\right)\begin{bmatrix}f_i^0\\f_i^1\end{bmatrix} = 0.$$

$\square$

2.2.3 First Order Collocation Method

So far, Galerkin-type methods are considered. In next two sections, our attention will be turned to collocation methods. In this subsection, we will construct a first order collocation scheme for (12). We start with introducing one collocation point $\xi_i \in I_i$ from each cell, and the actual choices will be specified later. A collocation method of the first order is then defined by requiring the piecewise constant numerical solution $f_h(k, t) \in V_h = V_h^0$ satisfy

$$\frac{\partial f_h(\xi_i, t)}{\partial t} = \hat{S}[f_h](\xi_i, t), \qquad \forall i \in \mathcal{N}. \tag{47}$$

We define $f_h(\xi_i, t) = f_i(t)$. Recall from Section 2.2.1,

$$f_h(\mathcal{K}_\alpha(E), t) + f_h(-\mathcal{K}_\alpha(E), t) = \sum_{j \in \mathcal{N}} \chi_{\Omega_{|j|}}(E) f_j(t) \tag{48}$$

then the scheme becomes

$$\frac{d}{dt} f_i(t) = -2\lambda_i f_i(t) + \sum_{j \in \mathcal{N}} s_{i,j} f_j(t). \tag{49}$$

Here,

$$\begin{aligned}
s_{i,j} = {}& \left(\frac{s_1(E, E + \varepsilon_p)}{\mathcal{K}_\alpha(E)/(1 + 2\alpha E)} \chi_{\Omega_{|j|}}(E) \right) \Big|_{E = E(\xi_i) - \varepsilon_p} \\
& + \left(\frac{s_{-1}(E, E - \varepsilon_p)}{\mathcal{K}_\alpha(E)/(1 + 2\alpha E)} \chi_{\Omega_{|j|}}(E) \right) \Big|_{E = E(\xi_i) + \varepsilon_p} \\
& + \left(\frac{s_0(E, E)}{\mathcal{K}_\alpha(E)/(1 + 2\alpha E)} \chi_{\Omega_{|j|}}(E) \right) \Big|_{E = E(\xi_i)},
\end{aligned} \tag{50}$$

and

$$\begin{aligned}
\lambda_i = {}& \frac{s_1(E - \varepsilon_p, E)}{\mathcal{K}_\alpha(E)/(1 + 2\alpha E)} \Big|_{E = E(\xi_i) + \varepsilon_p} + \frac{s_{-1}(E + \varepsilon_p, E)}{\mathcal{K}_\alpha(E)/(1 + 2\alpha E)} \Big|_{E = E(\xi_i) - \varepsilon_p} \\
& + \frac{s_0(E, E)}{\mathcal{K}_\alpha(E)/(1 + 2\alpha E)} \Big|_{E = E(\xi_i)}.
\end{aligned} \tag{51}$$

Again, the terms involving $E = E(\xi_i) - \varepsilon_p < 0$ are excluded.

Note that all the coefficients in the linear algebraic equation (49) are nonnegative, more specifically,

$$s_{i,j} \geq 0, \quad \lambda_i > 0, \qquad \forall i, j \in \mathcal{N}. \tag{52}$$

If we further require the collocation points are chosen to satisfy

$$\xi_i = -\xi_{-i}, \quad i \in \mathcal{N}^+, \tag{53}$$

then the energy function $E(k)$ being an even function implies $E(\xi_i) = E(\xi_{-i})$, and the following symmetries hold

$$\lambda_i = \lambda_{-i}, \quad s_{i,j} = s_{-i,j} = s_{i,-j} = s_{-i,-j}, \qquad \forall i, j \in \mathcal{N}.$$

Now we let

$$\Lambda = \text{diag}\{\lambda_1, \cdots, \lambda_N\}, \quad S = (s_{i,j})_{i,j \in \mathcal{N}^+}, \tag{54}$$

and $\mathbf{f}_- = [f_{-1}, \cdots, f_{-N}]^T, \mathbf{f}_+ = [f_1, \cdots, f_N]^T$, then the proposed scheme in (49) can be written as

$$\frac{d}{dt}\begin{bmatrix} \mathbf{f}_- \\ \mathbf{f}_+ \end{bmatrix} = \mathbb{S}\begin{bmatrix} \mathbf{f}_- \\ \mathbf{f}_+ \end{bmatrix} = \left(-2\begin{bmatrix} \Lambda & 0 \\ 0 & \Lambda \end{bmatrix} + \begin{bmatrix} S & S \\ S & S \end{bmatrix} \right)\begin{bmatrix} \mathbf{f}_- \\ \mathbf{f}_+ \end{bmatrix}. \tag{55}$$

Note $\mathbf{f}_+$ and $\mathbf{f}_-$ are defined differently from those in Section 2.2.1 and they do not contain the mesh parameter $\{\Delta k_i\}_i$.

If we further define $\mathbf{g} = (\mathbf{f}_- + \mathbf{f}_+)/2, \mathbf{h} = (\mathbf{f}_+ - \mathbf{f}_-)/2$, and $M = 2(S - \Lambda)$, then the proposed scheme (55) can be decoupled into two systems of halved size

$$\frac{d}{dt}\mathbf{g} = M\mathbf{g}, \quad \frac{d}{dt}\mathbf{h} = -2\Lambda\mathbf{h}. \tag{56}$$

Just as for the DG methods in Sections 2.2.1 and 2.2.2, if we are only concerned with the properties of the scattering matrix $\mathbb{S}$ regarding the discrete equilibrium, it is sufficient to simply consider $\frac{d}{dt}\mathbf{g} = M\mathbf{g}$.

Remark 2.7 Compared with Galerkin methods in Sections 2.2.1 and 2.2.2, collocation methods proposed here and in next subsection are much simpler to formulated and to implement. On the other hand, collocation methods in general do not preserve mass conservation property.

2.2.4   Fourier-Collocation Spectral Method

In this subsection, we will formulate a Fourier-collocation spectral method for the linear kinetic model with a singular scattering kernel, which is now reformulated into (12). It is assumed that the solution $f(k, t)$ is zero outside the interval $[-K_{\max}, K_{\max}]$, thus can be extended periodically. For simplicity, we use $K = K_{\max}$ throughout this subsection.

We seek an approximating solution $f_N(k, t)$ in the space $\hat{B}_N[-K, K] = \text{span}\{e^{i\frac{\pi}{K}nk}\}_{|n| \leq N}$, i.e.,

$$f_N(k, t) = \sum_{n=-N}^{N} \hat{f}_n(t) e^{i\frac{\pi}{K} nk}, \tag{57}$$

with the unknown coefficients $\hat{f}_n(t), n = -N, \cdots, N$, to be determined. For any function $g \in \hat{B}_N[-K, K]$, one can define its residual associated with the equation (1)

$$R_N(k, t; g) = \frac{\partial g(k, t)}{\partial t} - \hat{S}[g](k, t) = \frac{\partial g(k, t)}{\partial t} - \sum_{m=1}^{4} \mathcal{R}_m[g](k, t).$$

In the Fourier-collocation method, we require that the residual of the numerical solution $f_N(k, t)$ vanishes at a set of collocation grid points $\{k_j\}_{-N \le j \le N}$, defined as

$$k_j = K \frac{2j}{2N + 1}, \qquad -N \le j \le N.$$

Having this choice of the collocation points, the Fourier coefficients $\hat{f}_n(t)$ of the numerical solution $f_N(k, t)$ can be approximated by the discrete Fourier coefficients $\tilde{f}_n(t)$ based on the trapezoidal rule,

$$\tilde{f}_n(t) = \frac{1}{2N + 1} \sum_{j=-N}^{N} f_N(k_j, t) e^{-i\frac{\pi}{K} nk_j}. \tag{58}$$

Thus, the numerical solution $f_N(k, t)$, as a trigonometric polynomial, can also be expressed as

$$f_N(k, t) = \sum_{j=-N}^{N} f_N(k_j, t) g_j(k), \tag{59}$$

where $g_j(k)$ $(-N \le j \le N)$ is the Lagrange interpolation polynomial, given as

$$g_j(k) = \frac{\sin\left(\frac{2N+1}{2}\frac{\pi}{K}(k - k_j)\right)}{(2N + 1)\sin\left(\frac{\pi}{2K}(k - k_j)\right)} \tag{60}$$

and satisfying $g_j(k_n) = \delta_{jn}$. Now the Fourier-collocation method can be stated as follows. Look for $f_N(k, t)$ in the form of (59), such that

$$R_N(k_j, t; f_N) = \frac{\partial f_N(k_j, t)}{\partial t} - \hat{S}[f_N](k_j, t)$$

$$= \frac{\partial f_N(k_j, t)}{\partial t} - \sum_{m=1}^{4} \mathcal{R}_m[f_N](k_j, t) = 0, \qquad -N \le j \le N. \tag{61}$$

This yields $2N + 1$ equations to determine the $2N + 1$ point values $f_N(k_j, t)$, $j = -N, \cdots, N$, of the numerical solution.

Next we will convert the scheme to its algebraic form. From (12),

$$
\begin{aligned}
\mathscr{R}_1[f_N](k_j, t) &= \frac{s_1(E, E + \varepsilon_p)}{\mathscr{K}_\alpha(E)/(1 + 2\alpha E)} \Big(f(\mathscr{K}_\alpha(E), t) + f(-\mathscr{K}_\alpha(E), t) \Big)\big|_{E=E(k_j)-\varepsilon_p} \\
&= \frac{s_1(E, E + \varepsilon_p)}{\mathscr{K}_\alpha(E)/(1 + 2\alpha E)} \sum_{j=-N}^{N} f_N(k_j, t) \Big(g_j(\mathscr{K}_\alpha(E)) + g_j(-\mathscr{K}_\alpha(E)) \Big)\big|_{E=E(k_j)-\varepsilon_p}.
\end{aligned}
\tag{62}
$$

The remaining terms in (61) can be treated similarly. We define the solution vector $\mathbf{f}$ by collecting all the unknown coefficients in (59),

$$
\mathbf{f} = [f_N(k_{-N}, t), \ldots, f_N(k_{-1}, t), f_N(k_0, t), f_N(k_1, t), \ldots, f_N(k_N, t)]^\top \in \mathbb{R}^{2N+1},
$$

then the proposed Fourier-collocation method becomes a linear system

$$
\frac{d\mathbf{f}}{dt} = \mathbb{S}\mathbf{f},
\tag{63}
$$

where $\mathbb{S} = -2\Lambda + S \in \mathbb{R}^{(2N+1)\times(2N+1)}$, with

$$
\Lambda = \mathrm{diag}\{\lambda_{-N}, \cdots, \lambda_N\}, \qquad S = \big(s_{n,j}\big)_{n, j \in \{-N, \cdots, N\}},
\tag{64}
$$

and

$$
\begin{aligned}
\lambda_n &= \frac{s_1(E - \varepsilon_p, E)}{\mathscr{K}_\alpha(E)/(1 + 2\alpha E)}\big|_{E=E(k_n)+\varepsilon_p} + \frac{s_{-1}(E + \varepsilon_p, E)}{\mathscr{K}_\alpha(E)/(1 + 2\alpha E)}\big|_{E=E(k_n)-\varepsilon_p} \\
&\quad + \frac{s_0(E, E)}{\mathscr{K}_\alpha(E)/(1 + 2\alpha E)}\big|_{E=E(k_n)},
\end{aligned}
\tag{65}
$$

$$
\begin{aligned}
s_{n,j} &= \frac{s_1(E, E + \varepsilon_p)}{\mathscr{K}_\alpha(E)/(1 + 2\alpha E)} \Big(g_j(\mathscr{K}_\alpha(E)) + g_j(-\mathscr{K}_\alpha(E)) \Big)\big|_{E=E(k_n)-\varepsilon_p} \\
&\quad + \frac{s_{-1}(E, E - \varepsilon_p)}{\mathscr{K}_\alpha(E)/(1 + 2\alpha E)} \Big(g_j(\mathscr{K}_\alpha(E)) + g_j(-\mathscr{K}_\alpha(E)) \Big)\big|_{E=E(k_n)+\varepsilon_p} \\
&\quad + \frac{s_0(E, E)}{\mathscr{K}_\alpha(E)/(1 + 2\alpha E)} \Big(g_j(\mathscr{K}_\alpha(E)) + g_j(-\mathscr{K}_\alpha(E)) \Big)\big|_{E=E(k_n)},
\end{aligned}
\tag{66}
$$

and $n, j = -N, \cdots, N$. Given the notation is self-explained, the negative subindices are used for the entry of S for simplicity.

Similar to other collocation methods, our Fourier-collocation scheme does not satisfy mass conservation property. In terms of approximating the equilibrium of the scattering operator, this spectral method performs quite differently from the other methods in previous sections, see numerical examples in Section 2.3.4.

2.3 Numerical Experiments

In this section, we will demonstrate the performance of the numerical schemes when they are applied to two examples with the following parameter choices.

- **Parameter choice 1**. We consider the parabolic energy band model with $\alpha = 0$ in (3). There is no elastic collision, that is, $s_0 = 0$. In addition, we take the phonon energy $\varepsilon_p = 0.1$, lattice temperature $T = 0.0883$, transfer rate parameter $s_{-1} = 1$, and the maximum energy $E_{max} = 8$.
- **Parameter choice 2**. We use the nondimensionalized parameters for silicon, and this involves $\alpha = 0.01292$ in the energy model (3), phonon energy $\varepsilon_p = 2.43723$, lattice temperature $T = 1$, transfer rates $s_0 = 0.26531$ and $s_{-1} = 0.04432$, and the maximum energy $E_{max} = 16$.

Throughout, μ_j is the eigenvalue of M which has the jth largest real part, $j = 1, 2, \cdots$.

2.3.1 First Order Discontinuous Galerkin Method

In this section, we shall verify the results of the first order DG method. Notice that this method has also been studied numerically in [7] for the parabolic energy band model without the elastic term.

We use a uniform grid in the energy space with cell size ΔE. The method (25) is implemented with backward Euler method applied in time and $\Delta t = \Delta E$. The initial data are randomly generated, and it is nonnegative and normalized to have the same total mass as the exact equilibrium in (3). The criteria for stopping the time evolution is set to be $||\mathbf{g}^{old} - \mathbf{g}^{new}||_2, ||\mathbf{h}^{old} - \mathbf{h}^{new}||_2 \leq 10^{-7}$. The entries of S are computed using a midpoint rule quadrature, while the entries of Λ are obtained based on the column sum of M being zero.

Figures 1 and 2 contain comparison of one exact equilibrium and the computed equilibrium based on parameter choices 1 and 2. Here and in all the figures throughout Section 2.3, the exact equilibrium is taken as $f^e(k) = cf^G(k)$, where the normalized constant c is chosen so as to achieve the same total mass as the numerical solution. Figure 1 agrees well with the theory obtained in [7] (also see Section 2.2.1). When $\Delta E = \varepsilon_p/n$, and n is not an integer or $n = 1$, the matrix is irreducible, making the computed equilibrium closer to $f^G(k)$ qualitatively. When $n > 1$ is an integer, the scattering matrix M (hence $\mathbb{S}$) is reducible, and the dimension of its null space, also called kernel space, is bigger than one, specifically, $dim(ker(M)) = n$. In this

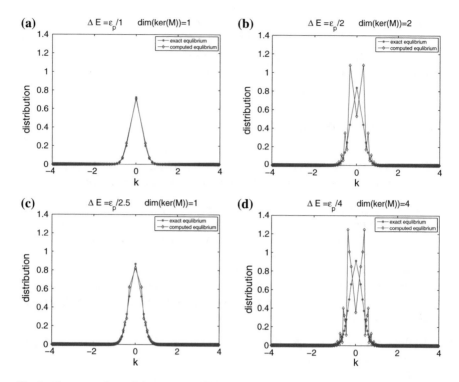

Fig. 1 The comparison of the exact equilibrium and the computed equilibrium by DG method with P^0 discrete space. The computed equilibrium is obtained by the backward Euler method with random initial data on uniform mesh in E with the indicated mesh size. Here and in all the figures throughout Section 2.3, the exact equilibrium is taken as $f^e(k) = cf^G(k)$ with some normalized constant c. Parameter choice 1. (a) $\Delta E = \varepsilon_p$. (b) $\Delta E = \varepsilon_p/2$. (c) $\Delta E = \varepsilon_p/2.5$. (d) $\Delta E = \varepsilon_p/4$.

case, the computed equilibrium distribution is no longer monotone in each half of the domain. Just as observed in Figure 2 of [7], each monotone subregion of the computed equilibrium involves n points on the grid, and this implies that the computed equilibrium is approximately in the form of $\hat{h}(E(k))\widehat{f^G}(k)$, where $\hat{h}(E)$ is approximating a ε_p-periodic grid-based function defined on the mesh grid of the energy domain, and $\widehat{f^G}(k)$ is an approximation for $f^G(k)$. In other words, the computed equilibrium captures the characteristics of the exact equilibriums. Computations based on parameter choice 2, which uses the Kane energy band model and has elastic scattering, as demonstrated in Figure 2 give a similar conclusion, verifying our claim in Section 2.2.1 and the results in [11].

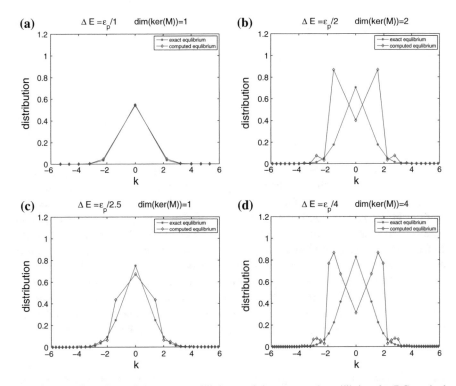

Fig. 2 The comparison of the exact equilibrium and the computed equilibrium by DG method with P^0 discrete space. The computed equilibrium is obtained by the backward Euler method with random initial data on uniform mesh in E with the indicated mesh size. Parameter choice 2. (a) $\Delta E = \varepsilon_p$. (b) $\Delta E = \varepsilon_p/2$. (c) $\Delta E = \varepsilon_p/2.5$. (d) $\Delta E = \varepsilon_p/4$.

2.3.2   Second Order Discontinuous Galerkin Method

In this subsection, we will present numerical experiments with the DG method using the P^1 discrete space introduced in Section 2.2.2. Particularly, we will investigate the importance of sufficiently accurate numerical quadratures, the dimension of $ker(M)$, and the accuracy of the scheme.

A close examination reveals that the integrals for computing the entries of Λ and S involve $E^{-1/2}$-type singularity near $E = 0$. In our implementation, the following strategy is adopted to compute $S_{i,j}, \Lambda_j$: When $j \leq n_{\text{singular}}$, we apply a special sixth order quadrature, obtained from the trapezoidal rule with Alpert correction to the left end of the reference element [1]; When $j > n_{\text{singular}}$, the standard 5-point Gauss quadrature is applied. To illustrate the effect of numerical quadratures, we consider the method implemented on a uniform mesh in k and $\Delta k = K_{\text{max}}/N$. The first 3 eigenvalues $\mu_{1,2,3}$ with the largest real part are reported in Table 1 for $N = 80$, and $n_{\text{singular}} = N/8, 2N/8, 3N/8$ and $4N/8$ with parameter choice 2. One can see that numerical quadratures with sufficiently large n_{singular} ensures that μ_1 is an accurate

Table 1 Effect of numerical quadrature by taking different values of n_{singular}. Uniform mesh in k with $\Delta k = K_{\max}/N$ and $N = 80$. Parameter choice 2.

n_{singular}	μ_1	$\mu_{2,3}$
$N/8, 2N/8$	1.31e-04	$-1.02\text{e-}08 \pm 9.39\text{e-}09i$
$3N/8$	2.27e-12	$-1.69\text{e-}08 \pm 1.30\text{e-}08i$
$4N/8$	4.33e-13	$-1.69\text{e-}08 \pm 1.30\text{e-}08i$

approximation for the zero eigenvalue, instead of contributing to a nontrivial growing mode. We further march the scheme with the equilibrium in (4) as the initial data and trapezoidal method in time with $\Delta t = \Delta k$, and plot in Figure 3 the numerical equilibriums compared with the exact one (again given by (4)) at time $t = 7$. The results confirm again the importance of using accurate enough numerical quadratures. In fact with n_{singular} being large enough, the numerical eigenvalues, other than those approximating zero, always have negative real part.

Next we examine how well our scheme approximates the dimension of $ker(M)$. Motivated by the P^0 results, we implement our DG method with the P^1 space on uniform meshes in E and $\Delta E = \varepsilon_p/n$. Both parameter choices are examined, with $n_{\text{singular}} = N/8$ for parameter choice 1 and $n_{\text{singular}} = 3N/8$ for parameter choice 2.

Fig. 3 Effect of numerical integration. Uniform mesh in k with $\Delta k = K_{\max}/N$ and $N = 80$. Trapezoidal method in time with $\Delta t = \Delta k$. Initial condition is the exact equilibrium in (4) with parameter choice 2.
(a) $n_{\text{singular}} = N/4$.
(b) $n_{\text{singular}} = 3N/8$.

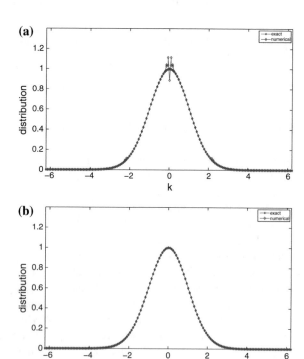

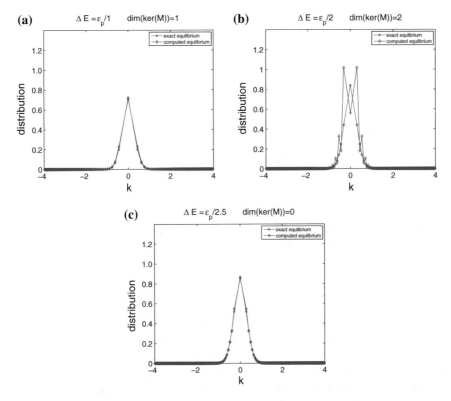

Fig. 4 The comparison of the exact equilibrium and the computed equilibrium by DG method with P^1 discrete space. The computed equilibrium is obtained by the backward Euler method with random initial data on uniform meshes in E with the indicated mesh size. Parameter choice 1 and $n_{singular} = N/8$. (a) $\Delta E = \varepsilon_p$. (b) $\Delta E = \varepsilon_p/2$. (c) $\Delta E = \varepsilon_p/2.5$.

When $n = m$ is an integer, the dimension of $ker(M)$ is m; and when $n = m + \frac{1}{2}$, the *numerical* dimension of $ker(M)$ is 1 in the sense that $\mu_1 = O(10^{-12,-13})$ and $\mu_2 = O(10^{-3,-5})$. This has been tested for $m = 1, \cdots, 10$. In Figure 4, we also plot the numerical equilibrium computed from marching the scheme in time with backward Euler method and $\Delta t = \Delta E, n = 1, 2, 2.5$. (When $n = 2.5$, the numerical dimension of $ker(M)$ is 1.) Though there is no mathematical analysis available, our numerical results seem to imply that the dependence of the (numerical) dimension of $ker(M)$ on the choice of the mesh grid in E for the DG method with the P^1 space is similar to that with the P^0 space. The computed equilibrium also shows the characteristics in (6) of the exact equilibriums. The setup for initialization and the stopping criteria is taken the same as in Section 2.3.1.

Finally, we turn to the accuracy of the scheme. In Table 2, we report the L_2 errors and convergence orders of the method at a fixed time $t = 7$ for both parameter choices. Uniform meshes in k are considered with $\Delta k = K_{max}/N$, and the initial condition is taken to be the exact equilibrium in (4). Second order accuracy is confirmed. In

Table 2 Accuracy of the DG method with P^1 discrete space at $t = 7$. Uniform mesh in k with $\Delta k = K_{\max}/N$. Trapezoidal method in time with $\Delta t = \Delta k$. Initial condition is the exact equilibrium in (4). $n_{\text{singular}} = N/8$ for parameter choice 1 and $n_{\text{singular}} = 3N/8$ for parameter choice 2.

	Parameter choice 1			Parameter choice 2		
N	L_2 error	order	μ_1	L_2 error	order	μ_1
40	4.83e-03	–	2.7125e-08	2.67e-03	–	6.0254e-10
80	1.81e-03	1.42	4.4021e-09	7.56e-04	1.82	2.2689e-12
160	4.68e-04	1.95	4.1503e-11	1.91e-04	1.98	−1.3997e-13
320	1.23e-04	1.92	−1.3055e-12	4.73e-05	2.02	−8.9108e-15
640	3.01e-05	2.03	−4.4868e-14	1.22e-05	1.95	–

addition, the leading eigenvalue μ_1 of M is also reported. Although this eigenvalue is not always negative, it converges to the zero eigenvalue as meshes are refined.

2.3.3 First Order Collocation Method

In this subsection, we will perform numerical study of the first order collocation method as outlined in Section 2.2.3. We compute the equilibrium using the backward Euler method, random initial data and stopping criteria $||\mathbf{g}^{\text{old}} - \mathbf{g}^{\text{new}}||_2$, $||\mathbf{h}^{\text{old}} - \mathbf{h}^{\text{new}}||_2 \leq 10^{-7}$. We consider both parameter choices 1 and 2 on uniform meshes in E or k. Since the collocation method does not achieve mass conservation, all computed equilibrium has been rescaled so that $\sum_i f_i \Delta k_i$ agrees with the exact equilibrium. To investigate the detailed performance of the method, we also obtain the leading eigenvalues of the scattering matrix M.

Figures 5 to 7 contain simulation results with parameter choice 1 on uniform meshes in E. The collocation points $\{\xi_i\}$ are chosen such that they correspond to midpoints in the computational grid for the energy variable. In particular, Figure 5 plots the results when $\Delta E = \varepsilon_p/n$, when n is an integer, while Figure 6 plots the solutions when n is not an integer. When compared with the first order Galerkin method, we can see that the results are similar when n is an integer, i.e., any integer $n > 1$ will yield the dimension of the kernel of the scattering matrix M to be bigger than one, producing oscillatory numerical equilibriums. However, the main difference occurs when n is a noninteger. From Figure 6, we can see when $n = 1.7, 2.2, 2.7, 3.2$, unlike DG method P^0 case, the collocation method still have $dim(ker(M)) > 1$. When $n = 2.5$, we can observe even from Figure 7(d) that the scattering matrix has several positive eigenvalues, which makes the time evolution scheme not converge to a steady state. Preliminary numerical tests show similar conclusions when $n = 1.5, 2.5, 3.5 \ldots 10.5$. From our numerical tests, it seems that if $n \in (N_n - 0.5, N_n + 0.5)$, where N_n is an integer, then $dim(ker(M)) = N_n$. We believe the different behavior of the collocation method when compared with the Galerkin method is because of the point-based

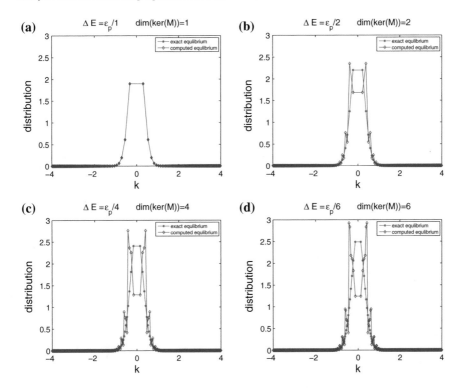

Fig. 5 The comparison of the exact equilibrium and the computed equilibrium by first order collocation method. The computed equilibrium is obtained by the backward Euler method with random initial data on uniform mesh in E with the indicated mesh size. Parameter choice 1. (a) $\Delta E = \varepsilon_p$. (b) $\Delta E = \varepsilon_p/2$. (c) $\Delta E = \varepsilon_p/4$. (d) $\Delta E = \varepsilon_p/6$.

nature of the collocation scheme. However, due to the lack of theoretical studies, we leave the detailed interpretation of this result to future work.

The next set of numerical tests was performed on uniform meshes in E with parameter choice 2. Figures 8 and 9 plot the equilibrium and the leading eigenvalues of the scattering matrix when $n = 1, 2, 1.7, 2.2$. Those selective results show $dim(ker(M)) = 0$ in all cases. However, for $n = 1.7, 2, 2.2$, there are two eigenvalues that are very close to zero, see Figure 8(e) for details. With the numerical dimension of $ker(M)$ being considered, the conclusion for parameter choice 2 is same as the one for parameter choice 1.

Finally, we plot the results for parameter choice 1 on uniform mesh in k when $N = 40, 80, 120, 160$ in Figure 10. The collocation points $\{\xi_i\}$ are chosen to be the midpoint in each cell in the k variable. Unlike the results for uniform mesh on E, the result for uniform mesh in k is not conclusive, i.e., this mesh choice does not imply the scattering matrix to be reducible/irreducible.

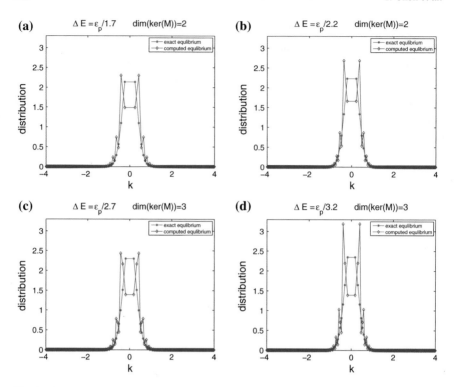

Fig. 6 The comparison of the exact equilibrium and the computed equilibrium by first order collocation method. The computed equilibrium is obtained by the backward Euler method with random initial data on uniform mesh in E with the indicated mesh size. Parameter choice 1. (a) $\Delta E = \varepsilon_p/1.7$. (b) $\Delta E = \varepsilon_p/2.2$. (c) $\Delta E = \varepsilon_p/2.7$. (d) $\Delta E = \varepsilon_p/3.2$.

In summary, the first order collocation method does not outperform the first order Galerkin scheme when measuring the qualitative behavior of the computed equilibrium. The collocation method, though being more computationally efficient, does not preserve mass conservation, and the results are highly dependent upon the choice of collocation points.

2.3.4 Fourier-Collocation Method

In this subsection, we will demonstrate the performance of the Fourier-collocation method defined in Section 2.2.4. This method behaves very differently from those we have examined so far, and it only captures the one-dimensional equilibrium $cf^G(k)$ (with $c > 0$ being a constant), given the computational domain is large enough. We attribute this to the global nature of this spectral method. By looking into the eigenvalues and corresponding eigenvectors of the scattering matrix $\mathbb{S}$ with parameter

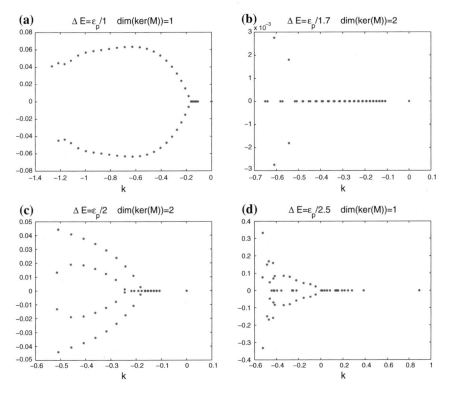

Fig. 7 The distribution of the first 50 eigenvalues of the scattering matrix by first order collocation method on uniform mesh in E with the indicated mesh size. Parameter choice 1. (a) $\Delta E = \varepsilon_p$. (b) $\Delta E = \varepsilon_p/1.7$. (c) $\Delta E = \varepsilon_p/2$. (d) $\Delta E = \varepsilon_p/2.5$.

choice 1 (in Table 3) and parameter choice 2 (in Table 4 and Table 5), the following observations can be made.

- **Parameter choice 1**

1. With the meshes being refined, the leading eigenvalue μ_1 is approaching 0 exponentially and $\mu_2 = O(10^{-3})$. When $N = 34$, this eigenvalue is zero at the round-off error level. The numerical dimension of the null space is one. In this case, we take $E_{max} = 8$ and, hence, $K_{max} = \mathscr{K}_\alpha(E_{max}) = 4$.
2. The eigenvector corresponding to μ_1 approximates the equilibrium $f^G(k)$. For comparison, the eigenvector is scaled such that the sum of its values at collocation points is the same as that of the exact equilibrium. In Table 3, we present errors of the computed eigenvector in l^∞, l^1 and l^2 vector norms. We also plot the scaled eigenvector with $N = 34$ in Figure 11, which captures the equilibrium with an error at the level of 10^{-13}.
3. The numerical equilibrium is also obtained by computing the steady state of the ODE system (63). The nonnegative initial data is chosen randomly, with

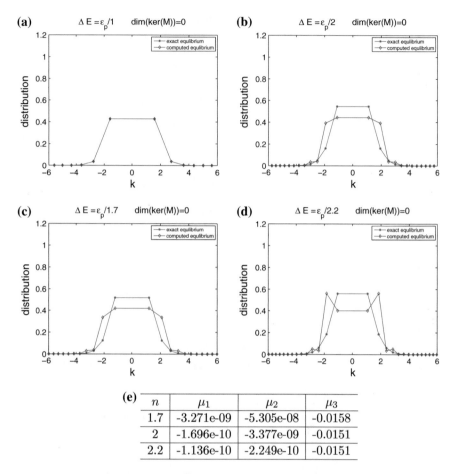

Fig. 8 The comparison of the exact equilibrium and the computed equilibrium. The computed equilibrium is obtained by the backward Euler method with random initial data on uniform mesh in E with the indicated mesh size. Parameter choice 2. (a) $\Delta E = \varepsilon_p$. (b) $\Delta E = \varepsilon_p/2$. (c) $\Delta E = \varepsilon_p/1.7$. (d) $\Delta E = \varepsilon_p/2.2$. (e) Leading eigenvalues of M : μ_1, μ_2, μ_3 are the eigenvalues with the three largest real part.

the stopping criteria as $\|\mathbf{f}^{old} - \mathbf{f}^{new}\|_\infty \le 10^{-10}$. In Figure 12, we compare the computed and the exact equilibrium. Though both the computed equilibriums before and after normalization well capture the shape of the equilibrium, the normalized one has a much smaller error at the level of 10^{-9}.

- **Parameter choice 2**

  We start with taking $E_{max} = 16$ in the computation.

1. Similar to parameter choice 1, the eigenvalue μ_1 of M is approaching 0 when N increases, with the convergence speed seemingly faster than that for parameter

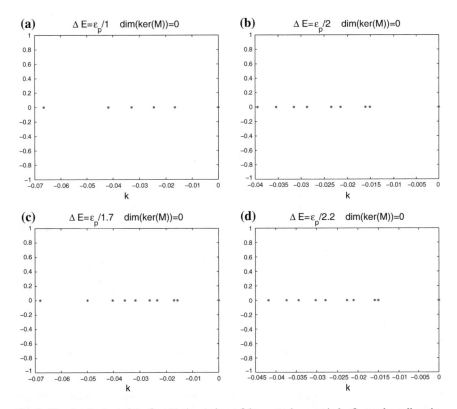

Fig. 9 The distribution of the first 11 eigenvalues of the scattering matrix by first order collocation method on uniform mesh in E with the indicated mesh size. Parameter choice 2. (a) $\Delta E = \varepsilon_p$. (b) $\Delta E = \varepsilon_p/2$. (c) $\Delta E = \varepsilon_p/1.7$. (d) $\Delta E = \varepsilon_p/2.2$.

choice 1. At the same time, μ_2 is $O(10^{-3,-4})$. The results in Table 4 are reported for N up to 10.

2. The eigenvector corresponding to μ_1 approximates the equilibrium $f^G(k)$. In Table 4, we report the errors between the scaled eigenvector and the exact equilibrium. For $N = 10$, the eigenvector approximates the equilibrium with an error at the level of 10^{-6}, as in Figure 13.

3. The numerical equilibrium is also obtained by computing the steady state of the ODE system (63). The nonnegative initial data are chosen randomly, with the stopping criteria as $\|\mathbf{f}^{old} - \mathbf{f}^{new}\|_\infty \leq 10^{-8}$. In Figure 14, we compare the computed and the exact equilibria with parameter choice 2 and $N = 10$. Again, the computed equilibrium after normalization has a smaller error at the level of 10^{-6}.

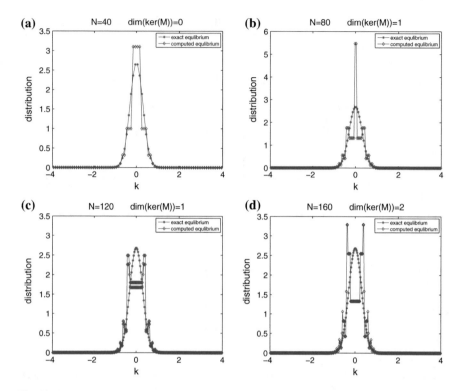

Fig. 10 The comparison of the exact equilibrium and the computed equilibrium. The computed equilibrium is obtained by the backward Euler method with random initial data on uniform mesh in k with the indicated mesh size. Parameter choice 1. (a) $N = 40$. (b) $N = 80$. (c) $N = 120$. (d) $N = 160$.

Table 3 Eigenvalues μ_1 and μ_2, together with the errors between the eigenvectors (normalized) corresponding to μ_1 and the exact equilibrium $f^G(k)$. Parameter choice 1.

N	$\mathrm{Re}(\mu_1)$	$\mathrm{Re}(\mu_2)$	Errors of the eigenvectors		
			l^∞	l^1	l^2
16	5.94e-04	−8.07e-03	9.73e-02	1.80e-02	2.79e-02
20	4.94e-05	−3.19e-03	3.16e-03	5.83e-04	9.03e-04
30	1.77e-11	−8.28e-03	1.78e-09	3.30e-10	5.11e-10
32	3.00e-13	−9.07e-03	4.88e-11	9.02e-12	1.40e-11
34	1.04e-15	−6.80e-03	5.55e-13	1.03e-13	1.60e-13

The results we have shown here are for N up to 10. For some larger values of N, it is observed that more than one computed eigenvalues of $\mathbb{S}$ can approach 0. There can also be multiple eigenvalues which have positive real parts. This is because the computational domain is not chosen large enough. To see this, we further test

Table 4 Eigenvalues μ_1 and μ_2, together with the errors between the eigenvectors (normalized) corresponding to μ_1 and the exact equilibrium $f^G(k)$. Parameter choice 2. $E_{max} = 16$.

N	$Re(\mu_1)$	$Re(\mu_2)$	Errors of the eigenvectors		
			l^∞	l^1	l^2
6	6.45e-2	−1.24e-03	6.43e-4	1.82e-2	2.38e-2
7	−1.19e-5	−3.36e-03	2.26e-3	8.89e-4	1.18e-3
8	7.12e-6	−3.89e-03	5.22e-4	1.29e-4	1.69e-4
9	−4.13e-7	−8.58e-04	7.73e-5	1.70e-5	2.47e-5
10	1.76e-8	−7.62e-04	1.54e-5	3.15e-6	5.42e-6

Table 5 Eigenvalues μ_1 and μ_2, together with the errors between the eigenvectors (normalized) corresponding to μ_1 and the exact equilibrium $f^G(k)$. Parameter choice 2. Larger domain size $E_{max} = 32$.

N	$Re(\mu_1)$	$Re(\mu_2)$	Errors of the eigenvectors		
			l^∞	l^1	l^2
11	−2.45e-05	−8.43e-04	1.37e-02	3.58e-03	4.91e-03
15	1.47e-07	3.60e-04	8.08e-05	2.30e-05	3.00e-05
20	−3.07e-12	1.49e-04	5.62e-08	6.89e-09	1.55e-08
22	9.58e-13	−1.34e-04	6.38e-09	8.58e-10	1.91e-09
27	1.40e-15	1.86e-05	7.73e-10	9.54e-11	2.22e-10

Fig. 11 The normalized eigenvector corresponding to μ_1 for $N = 34$ with exact equilibrium $f^G(k)$. Parameter choice 1.

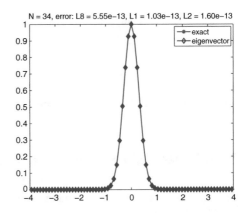

the method on a larger computational domain with $E_{max} = 32$. Again, the method captures the equilibrium $f^G(k)$ with the eigenvector corresponding to μ_1, as in Table 5 and Figure 15. With $N = 27$, μ_1 is $O(10^{-15})$, and the scaled eigenvector corresponding to μ_1 approximates $f^G(k)$ with an error at the level of 10^{-10}.

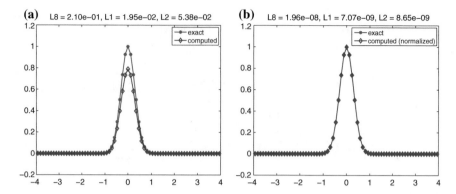

Fig. 12 The comparison of the exact equilibrium $f^G(k)$ and the computed equilibrium, obtained by the backward Euler method with random initial data and tolerance being $1.e - 10$. $N = 34$. Parameter choice 1. (a) Computed equilibrium before normalization. (b) Computed equilibrium after normalization.

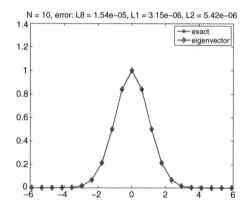

Fig. 13 The normalized eigenvector corresponding to μ_1 for $N = 10$ with exact equilibrium $f^G(k)$. Parameter choice 2. $E_{max} = 16$.

3 A Numerical Method for Continuous Scattering Kernels

In this section, we consider the kinetic model (1) with a continuous scattering kernel (7). If one follows the derivation in Section 2.2.1 to define a first order DG method for this model, it is easy to show that the scattering matrix is always irreducible when $\sigma(k, k') > 0$. Instead, we choose a different discretization which is well-suited for the model with a continuous scattering kernel.

Since the scattering kernel $S(k, k')$ has Gaussian decay and we are concerned with approximating the equilibrium solution, we assume there exists a constant K_{max} such that

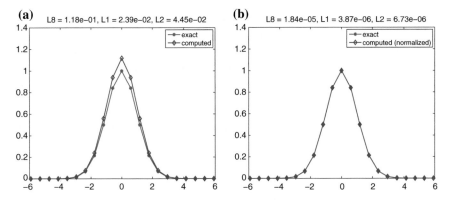

Fig. 14 The comparison of the exact equilibrium $f^G(k)$ and the computed equilibrium, obtained by the backward Euler method with random initial data and tolerance being $1.e-8$. $N = 10$. Parameter choice 2. $E_{max} = 16$. (a) Computed equilibrium before normalization. (b) Computed equilibrium after normalization.

Fig. 15 The normalized eigenvector corresponding to μ_1 for $N = 27$ with exact equilibrium $f^G(k)$. Parameter choice 2. Larger domain size $E_{max} = 32$.

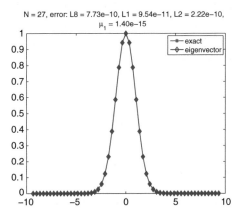

$$\left| \int_{-\infty}^{\infty} \left(S(k',k)f(k',t) - S(k,k')f(k,t) \right) dk' - \int_{-K_{max}}^{K_{max}} \left(S(k',k)f(k',t) - S(k,k')f(k,t) \right) dk' \right| < \varepsilon$$

for a user-prescribed tolerance ε for all k.

Equation (1) can now be discretized by applying numerical quadrature to the truncated domain. This technique is called *Nyström discretization* of the integral differential equation. Specifically, let $\{k_i\}_{i=1}^N$ denote the set of quadrature nodes in the interval $[-K_{max}, K_{max}]$ with corresponding weights $\{w_i\}_{i=1}^N$, then (1) is approximated by

$$\frac{\partial \hat{f}(k,t)}{\partial t} = \sum_{i=1}^{N} \left(S(k_i,k)\hat{f}(k_i,t) - S(k,k_i)\hat{f}(k,t) \right) w_i \tag{67}$$

where the solution $\hat{f}$ is an approximation to the exact solution f of (1). The quadrature points $\{k_i\}_{i=1}^N$ will be the discretization points.

To arrive at a linear system, the solution $\hat{f}$ is sought at the quadrature points k_j, $j = 1, \ldots, N$ for all t. The result is the following discrete ordinary differential equation

$$\frac{\partial \hat{f}(k_j, t)}{\partial t} = \sum_{i=1}^N \left(S(k_i, k_j) \hat{f}(k_i, t) - S(k_j, k_i) \hat{f}(k_j, t) \right) w_i \tag{68}$$

for each $j = 1, \ldots, N$ or in linear algebraic form

$$\frac{\partial \hat{f}}{\partial t}(t) = (\mathsf{S} - \Lambda) \hat{f}(t) = \mathsf{M} \hat{f}(t) \tag{69}$$

where $\mathsf{S}_{i,j} = S(k_j, k_i) w_j$, $\hat{f}$ denotes the vector of unknowns such that $\hat{f}_j = \hat{f}(k_j, t)$, $\Lambda = \text{diag}\{v\}$ and the vector v has entries given by $v_i = \sum_{j=1}^N S(k_i, k_j) w_j$.

Remark 3.1 Applying the numerical quadrature scheme to (68) (i.e., left multiplying (69) by w^T where $w_j = w_j$) shows that the discretization technique conserves total mass in time.

3.1 Numerical Experiments

The performance of the numerical method is explored in this section with two choices of $\sigma(k, k')$. In Section 3.1.1, the choice of σ results in a problem with a known solution while in Section 3.1.2 the choice of σ yields a problem without a reference solution.

For the numerical experiments, a ten-point composite Gaussian quadrature on equispaced panels is utilized to approximate the solution over the interval $[-K_{\max}, K_{\max}] = [-4, 4]$. Thus, the number of discretization points N is ten times the number of panels placed on the interval $[-4, 4]$.

3.1.1 An Example with a Known Solution

In this subsection, we illustrate the performance of the numerical method when $\sigma(k, k') = 1$. With this choice of σ, the exact solution is known to be $f_{\text{ex}}(k) = \frac{1}{\sqrt{2\pi}} e^{-k^2/2}$. Let f_{ex} denote the vector whose entries are f_{ex} evaluated at the discretization points.

Table 6 reports the number of discretization points N, the absolute error $E_{\text{abs}} = \|\hat{f} - f_{\text{ex}}\|_2$ and the relative error $E_{\text{rel}} = \frac{\|\hat{f} - f_{\text{ex}}\|_2}{\|f_{\text{ex}}\|_2}$ when computing the equilibrium solution, i.e., approximating solutions to (1) with $\frac{\partial f}{\partial t} = 0$. The numerical approximation is found by computing the null space of M in equation (69).

Table 6 The number of discretization points N, absolute error E_{abs}, and the relative error E_{rel} when applying the solution technique to equation (1) with $\sigma(k, k') = 1$.

N	E_{abs}	E_{rel}
10	9.49e-02	1.94e-01
20	1.32e-03	1.36e-03
40	4.13e-04	3.40e-04
80	1.43e-04	8.51e-05
160	5.05e-05	2.13e-05
320	1.79e-05	5.32e-06
640	6.32e-06	1.33e-06
1280	2.23e-06	3.33e-07
2560	7.90e-07	8.31e-08

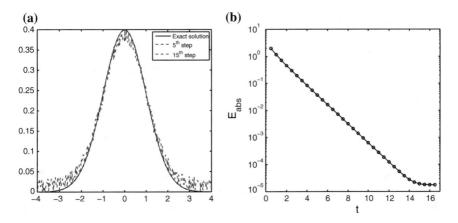

Fig. 16 (a) Approximate solutions after 5 and 10 time steps with a step size of $h = 0.5$. (b) Absolute error E_{abs} in approximate solution at time t.

Next, the backward Euler method was applied to (67) with a fixed $N = 320$ number of discretization points and timestep size $h = 0.5$. With this choice of N, Table 6 indicates that the expected converged accuracy should be approximately 1e-05. Thus, the iterative process is stopped when the norm of the difference between two iterates is less than 1e-05. Figure 16(a) illustrates the approximate solution at two different times in addition to the exact solution. Figure 16(b) illustrates the absolute error E_{abs} at each timestep. At the thirty-third timestep, the scheme has converged to the set tolerance.

Table 7 The number of discretization points N, absolute error E_{abs}, and the relative error E_{rel} when applying the solution technique to equation (1) with $\sigma(k, k') = (k - k')^2$.

N	E_{abs}	E_{rel}
10	2.36e-01	1.92e-01
20	2.47e-03	1.02e-03
40	7.77e-04	2.55e-04
80	2.69e-04	6.38e-05
160	9.50e-05	1.59e-05
320	3.36e-05	3.99e-06
640	1.19e-05	9.97e-07
1280	4.19e-06	2.49e-07

3.1.2 An Example with Unknown Solution

In this subsection, we consider (1) with $\sigma(k, k') = (k - k')^2$. For this choice of σ, the exact solution is unknown. In the first experiment, a convergence study is performed for the equilibrium problem. Let $\hat{f}_N$ denote the approximate solution obtained with N discretization points. Table 7 reports the number of discretization points N, the absolute convergence error $E_{abs} = \|\hat{f}_N - \mathsf{L}\hat{f}_{2N}\|_2$ where L is a matrix that interpolates $\hat{f}_{2N}$ at the $2N$ discretization points to the N coarse discretization points and the relative convergence error $E_{rel} = \frac{\|\hat{f}_N - \mathsf{L}\hat{f}_{2N}\|_2}{\|\mathsf{L}\hat{f}_{2N}\|_2}$.

Again backward Euler method is employed with timestep size $h = 0.5$ and $N = 320$ discretization points. We define the solution obtained by solving the equilibrium problem with $N = 320$ discretization points to be the reference solution.

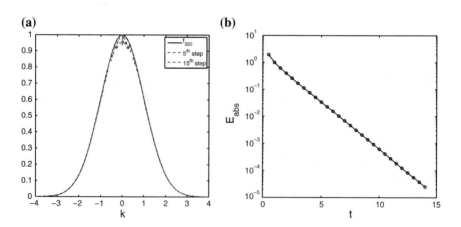

Fig. 17 (a) Approximate solutions after 5 and 10 timesteps with a step size of $h = 0.5$. (b) Absolute error E_{abs} in approximate solution at time t.

It takes 28 timesteps for the approximate solution to converge to the reference solution. Figure 17(a) illustrates the approximate solution after 5 and 15 timesteps. Figure 17(b) illustrates the absolute approximate error given by $E_{abs} = \| \hat{f}_{320} - \hat{f}(t) \|_2$ where $\hat{f}_{320}$ is the approximate equilibrium solution when $N = 320$ and $\hat{f}(t)$ is the approximate solution at time t.

4 Concluding Remarks

In this paper, we consider some one-dimensional space-homogeneous linear kinetic models arising from semiconductor device simulations. The focus of our efforts is to study the qualitative behaviors of the discrete scattering operators and the resulted numerical approximations for steady-state equilibrium. We review and discuss the mathematical results in [7, 11] for a first order finite volume method when it is applied to a model with δ-type singularity with the Kane energy band and the additional elastic scattering. Moreover, we investigate the numerical performance of first and higher order Galerkin method, a first order collocation method, and a Fourier-collocation spectral method for this model, as well as a Nyström method for a kinetic model with a continuous scattering kernel.

It seems to be nontrivial to generalize the analysis developed in [7, 11] to higher order and collocation-type schemes to solve models with δ-type singularity. For second (or higher) order Galerkin methods, the scattering matrix will become block structured, which requires additional tools in algebraic analysis. For collocation schemes, the analysis breaks down because the methods are no longer mass conservative. The numerical study in this paper seems to indicate that similar conclusion as for the discontinuous Galerkin scheme with the P^0 discrete space holds for the discontinuous Galerkin scheme with the P^1 space regarding how the properties of the kernel of the discrete scattering operator depend on the mesh choices. The first order collocation method computes numerical equilibrium that is highly dependent on the mesh, while the Fourier-collocation method, with its global nature, only captures a one-dimensional equilibrium associated with $f^G(k)$, and the resulting approximation is very accurate with the spectral accuracy of the method. These numerical results motivate our immediate future work on the theoretical analysis of some of the methods. Another interesting future direction consists of generalization to higher dimensions. Real-world applications call for attention to models in higher dimensions with transport effect. Such models have different equilibria from the space homogeneous case and the analysis will be more involved.

Acknowledgments The third author was partially supported by NSF Grant DMS-1318186, and the fifth author was partially supported by NSF Grants DMS-0847241 and DMS-1318409.

References

1. B. K. Alpert. Hybrid Gauss-trapezoidal quadrature rules. *SIAM Journal on Scientific Computing*, 20(5):1551–1584, 1999.
2. J. A. Carrillo, I. M. Gamba, A. Majorana, and C.-W. Shu. A WENO-solver for the transients of Boltzmann-Poisson system for semiconductor devices: performance and comparisons with Monte Carlo methods. *Journal of Computational Physics*, 184(2):498–525, 2003.
3. Y. Cheng, I. M. Gamba, A. Majorana, and C.-W. Shu. A discontinuous Galerkin solver for Boltzmann-Poisson systems in nano devices. *Computer Methods in Applied Mechanics and Engineering*, 198(37):3130–3150, 2009.
4. E. Fatemi and F. Odeh. Upwind finite difference solution of Boltzmann equation applied to electron transport in semiconductor devices. *Journal of Computational Physics*, 108(2):209–217, 1993.
5. D. K. Ferry. *Semiconductors*. IoP Publishing, 2013.
6. Y. L. Le Coz. *Semiconductor device simulation: a spectral method for solution of the Boltzmann transport equation*. PhD thesis, Massachusetts Institute of Technology, Dept. of Electrical Engineering and Computer Science, 1988.
7. R. Li, T. Lu, and W. Yao. Discrete kernel preserving model for 1d electron-optical phonon scattering. *Journal of Scientific Computing*, 62(2):317–335, 2015.
8. M. Lundstrom. *Fundamentals of carrier transport*. Cambridge University Press, 2009.
9. A. Majorana. Equilibrium solutions of the non-linear boltzmann equation for an electron gas in a semiconductor. *Il Nuovo Cimento B*, 108(8):871–877, 1993.
10. P. Markowich, C. Ringhofer, and C. Schmeiser. *Semiconductor equations*. Springer-Verlag, 1990.
11. W. Yao. *Simulation of one-dimensional semiconductors and eigen analysis of discrete electron optical phonon scattering*. PhD thesis, Peking University, School of Mathematical Sciences, 2014.

On Metrics for Computation of Strength of Coupling in Multiphysics Simulations

Anastasia Wilson, Wei Du, Guanglian Li, Azam Moosavi and Carol S. Woodward

Abstract Many multiphysics applications arise in the world of mathematical modeling and simulation. Much of the time in scientific computation these multiphysics applications are solved by decoupling the physics, giving no heed to how this affects the numerical results. However, a fully coupled approach is often not computationally cost effective. Consequently, having a metric for determining the strength of coupling could give insight into whether a simulation should be decoupled in the computation. If the fully coupled approach is not available, then a metric that measures the strength of coupling dynamically in time could help determine when smaller time steps are required to better incorporate coupling into the split solution. In this paper, we report on an Institute for Mathematics and Its Applications student project where we explored metrics for dynamically measuring the strength of coupling between two physical components in a model multiphysics simulation. Four metrics were considered: two based on measured components of the Jacobian matrix, one on error estimates, and the last on timescales of the system components. The metrics are

A. Wilson
Department of Mathematical Sciences, Clemson University, Clemson, SC, USA
e-mail: anastas@clemson.edu

W. Du
Institute for Computational Engineering and Sciences,
University of Texas at Austin, Austin, TX, USA
e-mail: weidu@ices.utexas.edu

G. Li
Institute for Numerical Simulation, University of Bonn, Bonn, Germany
e-mail: lotusli0707@gmail.com

A. Moosavi
Department of Computer Science,
Virginia Polytechnic Institute and State University, Blacksburg, VA, USA
e-mail: azmosavi@vt.edu

C.S. Woodward (✉)
Lawrence Livermore National Laboratory,
Center for Applied Scientific Computing, Livermore, CA, USA
e-mail: woodward6@llnl.gov

© Springer Science+Business Media New York 2016
S.C. Brenner (ed.), *Topics in Numerical Partial Differential Equations
and Scientific Computing*, The IMA Volumes in Mathematics
and its Applications 160, DOI 10.1007/978-1-4939-6399-7_6

all developed based on the previous work found in the literature and tested on a diffusion–reaction problem.

1 Introduction

In recent years, more attention has been paid to multiphysics modeling for better understanding of integrated processes. Multiphysics systems are often characterized by coupled systems with multiple time or length scales. Typically, the system consists of different types of partial differential equations, possibly even in different domains. Common applications include surface and subsurface hydrology, radiation hydrodynamics, geodynamics, and climate change models. In most cases, the system is coupled.

A common approach to these coupled systems is to apply an operator splitting method in order to make use of prior code development in the single physics components. In operator splitting techniques, the system of PDEs is decomposed, the simpler subproblems are solved individually, and the solutions are combined after each time step in a way to preserve accuracy, assuming the individual components are sufficiently independent of each other. The subproblems can be discretized independently and treated with different time steps. The advantage of splitting methods is in computational efficiency when the operators are only weakly coupled. However, these splitting approaches can shed accuracy and stability if the operators are tightly coupled [3]. In particular, the solution can diverge even when the problem is well defined for all time [1, 7]. This method also introduces an error related to the splitting scheme itself, independent of the methods used in each of the component systems.

The trade-off between using split schemes that are efficient for each component and a more expensive fully coupled scheme that does not suffer from stability constraints and does not have the splitting error can be large. In this paper, we report on a student project for an Institute for Mathematics and its Applications Workshop where our goal is to develop efficient and computable metrics to help determine when the physics components in a multiphysics system are tightly or weakly coupled. Since this strength of coupling can change in time as the simulation progresses, it is important that these metrics are cheap to compute as they will need to be evaluated at each time step. As a test model, we define and test metrics on a coupled diffusion–reaction system where the strength of coupling between the two physics components changes over time.

This paper examines four metrics for strength of coupling. The first two use information from a Jacobian of the coupled system. The third is based on the timescales of the system components. Lastly, we look at a metric derived from the error analysis of a decoupled model.

The paper is organized as follows. Section 2 casts the developed metrics for determining the strength of coupling. The numerical experiments focus on diffusion–reaction systems, and the test problems used for this system are defined in Section 3.

Numerical experiments to illustrate the proposed schemes are carried out in Section 4, and conclusions are furnished in Section 5.

2 Strength of Coupling Metrics

In generic form, we are interested in the investigation of the coupling strength of the following nonlinear equations:

$$\begin{cases} F_1(u, v; x, t) = 0 \\ F_2(u, v; x, t) = 0. \end{cases} \tag{1}$$

Here, $u = u(x, t)$ and $v = v(x, t)$ denote the unknowns, $x \in \mathbb{R}$ denotes the space variable, and t is the time variable. In this paper, the nonlinear equations above are derived from the discretization of a time-dependent diffusion–reaction problem. The metrics we consider could be extended to problems with more than two physical components. For clarity of exposition, however, we consider only two components in this work.

2.1 Norm of Off-Diagonal Jacobian Blocks

One way to determine how strongly u and v are coupled is by looking at the sensitivities of F_1 and F_2 to u and v [2], that is, by looking at the Jacobian:

$$J = \begin{bmatrix} J_{11} & J_{12} \\ J_{21} & J_{22} \end{bmatrix}$$

with

$$J_{11} = \frac{\partial F_1}{\partial u}, \ J_{12} = \frac{\partial F_1}{\partial v}, \ J_{21} = \frac{\partial F_2}{\partial u}, \ J_{22} = \frac{\partial F_2}{\partial v}.$$

Recalling the previous work from [2], once we have a converged solution from the nth time step, (u^n, v^n), then we can take a Newton iteration to get a first approximation for the solution at the $(n + 1)$st time step. In doing this, we have

$$\begin{bmatrix} J_{11}^{(n)} & J_{12}^{(n)} \\ J_{21}^{(n)} & J_{22}^{(n)} \end{bmatrix} \begin{bmatrix} \Delta u^{(n)} \\ \Delta v^{(n)} \end{bmatrix} = \begin{bmatrix} -F_1^{(n)} \\ -F_2^{(n)} \end{bmatrix}$$

$$\Rightarrow J_{11}^{(n)} \Delta u^{(n)} + J_{12}^{(n)} \Delta v^{(n)} = -F_1^{(n)}$$

$$\Rightarrow \Delta u^{(n)} = -\left(J_{11}^{(n)}\right)^{-1} J_{12}^{(n)} \Delta v^{(n)} - \left(J_{11}^{(n)}\right)^{-1} F_1^{(n)}$$

$$\Rightarrow J_{21}^{(n)} \left[-\left(J_{11}^{(n)}\right)^{-1} J_{12}^{(n)} \Delta v^{(n)} - \left(J_{11}^{(n)}\right)^{-1} F_1^{(n)} \right] + J_{22}^{(n)} \Delta v^{(n)} = -F_2^{(n)}$$

$$\Rightarrow \left[-J_{21}^{(n)} \left(J_{11}^{(n)}\right)^{-1} J_{12}^{(n)} + J_{22}^{(n)} \right] \Delta v^{(n)} = -F_2^{(n)} + J_{21}^{(n)} \left(J_{11}^{(n)}\right)^{-1} F_1^{(n)} \qquad (2)$$

Notice that if the coupling is one way, specifically if u does not depend on v, then $J_{12}^{(n)} \Delta v^{(n)}$ equals zero and (2) simplifies to

$$J_{22}^{(n)} \Delta v^{(n)} = -F_2^{(n)} + J_{21}^{(n)} \left(J_{11}^{(n)}\right)^{-1} F_1^{(n)},$$

which has the same left-hand side as if the problem is uncoupled. A similar situation arises in the case when v does not depend on u as that implies $J_{21}^{(n)} \Delta u^{(n)} = 0$. If we assume that $\Delta u^{(n)} \neq 0$ and $\Delta v^{(n)} \neq 0$, then we have either $J_{12}^{(n)} = 0$ or $J_{21}^{(n)} = 0$ in the cases of one-way or no coupling. Consequently, the off-diagonal blocks of the Jacobian matrix at the nth time step, $J_{12}^{(n)}$ and $J_{21}^{(n)}$, or more specifically the norms of the off-diagonal blocks at the nth time step, $\|J_{12}^{(n)}\|$ and $\|J_{21}^{(n)}\|$, may be able to be used as a metric to determine the strength of the coupling between u and v at the nth time step.

    This metric can be extremely insightful in the case of an explicit statement of the problem, i.e., if F_1 and F_2 are known, and the partial derivatives can be calculated. However, this is not necessarily the case in practical applications. In these cases, numerical approximations, such as finite differences, may be applied to approximate the derivatives. The accuracy and cost of a finite difference approximation would depend on the amount of information available about F_1 and F_2, but the approximation could potentially be computed using the residuals as described in [2]. Consequently, the main downfall of this metric is the wealth of information required to compute it exactly.

2.2 Condition Number of Diagonal Jacobian Blocks

We also observe that the condition number of the subproblems has an impact on convergence. From (2), we find that the term $J_{21} J_{11}^{-1} F_1$ on the right-hand side may greatly influence the solution. Let κ_{11} be the condition number of J_{11}. If κ_{11} is large, the inverse, J_{11}^{-1}, may not be computed or approximated accurately since

$$\|J_{11}^{-1}\| = \frac{\kappa_{11}}{\|J_{11}\|} \qquad (3)$$

Under the assumption κ_{11} is large, we may expect large perturbation to the right-hand side of (2). Therefore, the solution to the v subproblem depends on the condition number of the first diagonal block of the u-problem Jacobian matrix. In fact, this condition number is a coefficient on the current residual of the first subproblem. So, the amount to which the first subproblem is not satisfied factors into the solution of the second subproblem.

Consequently, if J_{21} is of sufficient size, the solution to (2) will mainly depend on κ_{11}. Similarly, the accuracy of the u subproblem will be a function of the condition number of J_{22}. Therefore, we can look into the condition number of both of the diagonal blocks of the Jacobian matrix and use the value in the worst of the subproblems as another way to determine when the coupling is strong.

This metric works nicely for the cases where J_{21} or J_{12} is equal to the identity matrix. But, for general forms of J_{21} or J_{12}, the efficiency of this metric is not clear. The factors of applicability and restrictions are still under exploration. Another thing that is worth mentioning is the computation of the condition number. Based on the definition in (3), this metric requires information about the inverse matrix, which could be expensive to compute for large problems. Thus, if this metric proves to be useful, approximations will be necessary to estimate the condition number.

2.3 Time Scales

Every physical phenomenon has a timescale at which it develops; consequently, multiphysics applications can have multiple time scales occurring simultaneously. Depending on how these scales interact with one another, the overall dynamic scale of the system can change [4–6]. When the timescales of the physics involved in F_1 and F_2 have vastly different values (orders of magnitude different), then the dynamic timescale and long-term behavior are fairly predictable; specifically, the dynamic timescale is expected to be on the order of the smallest physical timescale involved in the problem. However, in the case when component timescales approximately balance, the dynamic timescale and long-term behavior are harder to predict. In essence, the dynamic timescale (and overall end behavior) in this case is highly dependent on the interaction of the physics in the system. Therefore, determining at each time step if the timescales are comparable could be a possible metric to determine how strongly the physics are coupled.

Developing a metric based on timescales could be very beneficial as it would be computationally inexpensive. At the most, this metric would require some simple finite difference evaluations to approximate the timescales above. However, as these terms are already computed during most numerical solution processes, a metric based on timescales could be evaluated by simply reusing already computed values. The main downfall of a timescales metric is that there is no theoretical basis to support the efficacy of such a metric.

2.4 Error Matrix

In this section, we try to estimate the error between the fully coupled system and the decoupled system for (1) in the sense of the L^2 norm based on [2]. We restate the weakly coupled algorithm from [2] in Algorithm 1 for the sake of completeness.

Algorithm 1: Weak coupled algorithm to solve the system by breaking the coupling

---

Input: (u_0, v_0) and ϵ

for $i = 1, 2, \ldots$, *until* $\|\,\|F_1(u_{i-1}, v_{i-1})\| + \|F_2(u_{i-1}, v_{i-1})\|\,\| < \epsilon$ **do**
  Solve for u in $F_1(u, v_{i-1}) = 0$
  $u_i \leftarrow u$
  Solve for v in $F_2(u_i, v) = 0$
  $v_i \leftarrow v$
end

As to the reference solution (i.e., the fully coupled solution) of (1), we solve with a Newton iteration as stated in Section 2.1. Also, a Newton iteration is applied to solve the subproblems in Algorithm 1. For clarity of presentation, we assume the Newton iteration numbers solving for the first unknown and the second unknown are the same in Algorithm 1.

Take $(\hat{u}^0, \hat{v}^0)$ as the initial guess and $\hat{u}^{(l)}, \hat{v}^{(l)}$ as the lth iteration solution for solving u, v in the decoupling subproblems, respectively. Then, the Newton iteration for those two unknowns can be written as

$$\begin{bmatrix} J_{11}^{(l-1)} & 0 \\ 0 & J_{22}^{(l-1)} \end{bmatrix} \begin{bmatrix} \hat{u}^{(l)} - \hat{u}^{(l-1)} \\ \hat{v}^{(l)} - \hat{v}^{(l-1)} \end{bmatrix} = - \begin{bmatrix} F_1(\hat{u}^{(l-1)}, \hat{v}^0) \\ F_2(\hat{u}^1, \hat{v}^{(l-1)}) \end{bmatrix}. \tag{4}$$

Here, $\hat{u}^1$ denotes the convergent solution for the first unknown u.

Reorganizing the linear system above by constructing the Newton iteration for the coupled system, we see

$$\begin{bmatrix} J_{11}^{(l-1)} & J_{12}^{(l-1)} \\ J_{21}^{(l-1)} & J_{22}^{(l-1)} \end{bmatrix} \begin{bmatrix} \hat{u}^{(l)} - \hat{u}^{(l-1)} \\ \hat{v}^{(l)} - \hat{v}^{(l-1)} \end{bmatrix} = - \begin{bmatrix} F_1(\hat{u}^{(l-1)}, \hat{v}^{(l-1)}) \\ F_2(\hat{u}^{(l-1)}, \hat{v}^{(l-1)}) \end{bmatrix} + E_{rr}, \tag{5}$$

where

$$E_{rr} = \begin{bmatrix} 0 & J_{12}^{(l-1)} \\ J_{21}^{(l-1)} & 0 \end{bmatrix} \begin{bmatrix} \hat{u}^{(l)} - \hat{u}^{(l-1)} \\ \hat{v}^{(l)} - \hat{v}^{(l-1)} \end{bmatrix} + \begin{bmatrix} F_1(\hat{u}^{(l-1)}, \hat{v}^{(l-1)}) - F_1(\hat{u}^{(l-1)}, \hat{v}^0) \\ F_2(\hat{u}^{(l-1)}, \hat{v}^{(l-1)}) - F_2(\hat{u}^1, \hat{v}^{(l-1)}) \end{bmatrix}. \tag{6}$$

Now, we will estimate E_{rr}.

Using a Taylor expansion, we have

$$F_1(\hat{u}^{(l-1)}, \hat{v}^{(l-1)}) - F_1(\hat{u}^{(l-1)}, \hat{v}^0) = J_{12}^{(l-1)}(\hat{v}^{(l-1)} - \hat{v}^0) + o(|\hat{v}^{(l-1)} - \hat{v}^0|),$$

and

$$F_2(\hat{u}^{(l-1)}, \hat{v}^{(l-1)}) - F_2(\hat{u}^1, \hat{v}^{(l-1)}) = -J_{21}^{(l-1)}(\hat{u}^1 - \hat{u}^{l-1}) + o(|\hat{u}^{(l-1)} - \hat{u}^1|).$$

Therefore,

$$E_{rr} = \begin{bmatrix} 0 & J_{12}^{(l-1)} \\ J_{21}^{(l-1)} & 0 \end{bmatrix} \begin{bmatrix} \hat{u}^{(l)} - \hat{u}^1 \\ \hat{v}^{(l)} - \hat{v}^0 \end{bmatrix} + \begin{bmatrix} o(|\hat{v}^{(l-1)} - \hat{v}^0|) \\ o(|\hat{u}^{(l-1)} - \hat{u}^1|) \end{bmatrix}.$$

Substituting the result above into (5) yields,

$$\begin{bmatrix} J_{11}^{(l-1)} & J_{12}^{(l-1)} \\ J_{21}^{(l-1)} & J_{22}^{(l-1)} \end{bmatrix} \begin{bmatrix} \hat{u}^{(l)} - \hat{u}^{(l-1)} \\ \hat{v}^{(l)} - \hat{v}^{(l-1)} \end{bmatrix} = - \begin{bmatrix} F_1(\hat{u}^{(l-1)}, \hat{v}^{(l-1)}) \\ F_2(\hat{u}^{(l-1)}, \hat{v}^{(l-1)}) \end{bmatrix} + \begin{bmatrix} 0 & J_{12}^{(l-1)} \\ J_{21}^{(l-1)} & 0 \end{bmatrix} \begin{bmatrix} \hat{u}^{(l)} - \hat{u}^1 \\ \hat{v}^{(l)} - \hat{v}^0 \end{bmatrix}$$
$$+ \begin{bmatrix} o(|\hat{v}^{(l-1)} - \hat{v}^0|) \\ o(|\hat{u}^{(l-1)} - \hat{u}^1|) \end{bmatrix}.$$

Denote the inverse of $J = \begin{bmatrix} J_{11}^{(l-1)} & J_{12}^{(l-1)} \\ J_{21}^{(l-1)} & J_{22}^{(l-1)} \end{bmatrix}$ as J^{-1}. Then, we obtain,

$$\begin{bmatrix} \hat{u}^{(l)} - \hat{u}^{(l-1)} \\ \hat{v}^{(l)} - \hat{v}^{(l-1)} \end{bmatrix} = - J^{-1} \begin{bmatrix} F_1(\hat{u}^{(l-1)}, \hat{v}^{(l-1)}) \\ F_2(\hat{u}^{(l-1)}, \hat{v}^{(l-1)}) \end{bmatrix} + J^{-1} \begin{bmatrix} 0 & J_{12}^{(l-1)} \\ J_{21}^{(l-1)} & 0 \end{bmatrix} \begin{bmatrix} \hat{u}^{(l)} - \hat{u}^1 \\ \hat{v}^{(l)} - \hat{v}^0 \end{bmatrix}$$
$$+ \|J^{-1}\| \begin{bmatrix} o(|\hat{v}^{(l-1)} - \hat{v}^0|) \\ o(|\hat{u}^{(l-1)} - \hat{u}^1|) \end{bmatrix}.$$

Let L be the iteration number for both of the subproblems. Repeating the process above for each iteration, we obtain

$$\begin{bmatrix} \hat{u}^{(L)} \\ \hat{v}^{(L)} \end{bmatrix} = \begin{bmatrix} \hat{u}^0 \\ \hat{v}^0 \end{bmatrix} - \sum_{s=0}^{L-1} J_s^{-1} \begin{bmatrix} F_1(\hat{u}^{(s)}, \hat{v}^{(s)}) \\ F_2(\hat{u}^{(s)}, \hat{v}^{(s)}) \end{bmatrix} + \sum_{s=0}^{L-1} J_s^{-1} \begin{bmatrix} 0 & J_{12}^{(s)} \\ J_{21}^{(s)} & 0 \end{bmatrix} \begin{bmatrix} \hat{u}^{(s+1)} - \hat{u}^1 \\ \hat{v}^{(s+1)} - \hat{v}^0 \end{bmatrix}$$
$$+ \sum_{s=0}^{L-1} \|J_s^{-1}\| \begin{bmatrix} o(|\hat{v}^{(s)} - \hat{v}^0|) \\ o(|\hat{u}^{(s)} - \hat{u}^1|) \end{bmatrix}.$$

Here, J_s^{-1} denotes the inverse of the Jacobian matrix at iteration s.
Note that the Newton iteration for the coupled system is

$$\begin{bmatrix} \hat{u}^{(L)} \\ \hat{v}^{(L)} \end{bmatrix} = \begin{bmatrix} \hat{u}^0 \\ \hat{v}^0 \end{bmatrix} - \sum_{s=0}^{L-1} J_s^{-1} \begin{bmatrix} F_1(\hat{u}^{(s)}, \hat{v}^{(s)}) \\ F_2(\hat{u}^{(s)}, \hat{v}^{(s)}) \end{bmatrix}.$$

Thus, necessary conditions for the decoupled solution to be close to the coupled solution are that $||J^{-1}||$ is bounded and $\left\|J^{-1}\begin{bmatrix} 0 & J_{12}^{(l-1)} \\ J_{21}^{(l-1)} & 0 \end{bmatrix}\right\|$ is very small.

Remark 1 This result is consistent with Section 2.1. In that section, the off-diagonal Jacobian blocks J_{12} and J_{21} are used as the indicator for a stronger coupling or a weaker coupling.

As a result of this remark, we do not show any direct results with this metric.

3 Test System

We evaluate the strength of coupling metrics on a model diffusion–reaction system. In 1D, the problem is written as

$$\frac{\partial u}{\partial t} - d(t)\frac{\partial^2 u}{\partial x^2} = f(u, t) \quad x \in [x_0, x_1], t \in [t_0, t_1], \tag{7}$$

$$u(x_0, t) = a, \tag{8}$$

$$u(x_1, t) = b, \tag{9}$$

$$u(x, t_0) = u_0(x). \tag{10}$$

For the fully coupled model, we discretize in space with a finite difference method and in time with Backward Euler. The discrete form is

$$\frac{u_j^{n+1} - u_j^n}{\Delta t} - d(t^{n+1})\frac{u_{j+1}^{n+1} - 2u_j^{n+1} + u_{j-1}^{n+1}}{\Delta x^2} = f(u_j^{n+1}, t^{n+1}). \tag{11}$$

This equation is equivalent to the system

$$F_j(u_j^{n+1}, u_{j-1}^{n+1}, u_{j+1}^{n+1}) = -d(t^{n+1})u_{j-1}^{n+1} + (\Delta x^2/\Delta t + 2d(t^{n+1}))u_j^{n+1}$$
$$- \Delta x^2 f(u_j^{n+1}) - d(t^{n+1})u_{j+1}^{n+1} - \Delta x^2 u_j^n/\Delta t = 0. \tag{12}$$

We use Newton's method to solve. The Jacobian matrix, J, is represented as $\{J_{ij}\} = \left\{\frac{\partial F_i}{\partial u_j}\right\}$. Using the finite difference spatial discretization with a uniform grid of m discretization points and a backward Euler temporal discretization with N uniform temporal discretization points, we obtain the tridiagonal matrix, J, as follows:

$$J_{jj} = \frac{\partial F_j}{\partial u_j} = \frac{\Delta x^2}{\Delta t} + 2d(t^{n+1}) - \Delta x^2 \frac{\partial f}{\partial u_j}, \tag{13}$$

$$J_{j,j-1} = \frac{\partial F_j}{\partial u_{j-1}} = -\Delta x^2 d(t^{n+1}), \tag{14}$$

$$J_{j-1,j} = \frac{\partial F_j}{\partial u_{j+1}} = -\Delta x^2 d(t^{n+1}), \tag{15}$$

for $1 < j < m$. When $j = 1$,

$$J_{11} = \frac{\partial F_1}{\partial u_1} = \frac{\Delta x^2}{\Delta t} + d(t^{n+1}) - \Delta x^2 \frac{\partial f}{\partial u_1}, \tag{16}$$

$$J_{12} = \frac{\partial F_1}{\partial u_2} = -d(t^{n+1}), \tag{17}$$

and when $j = m$

$$J_{mm} = \frac{\partial F_m}{\partial u_m} = \frac{\Delta x^2}{\Delta t} + d(t^{n+1}) - \Delta x^2 \frac{\partial f}{\partial u_m}, \tag{18}$$

$$J_{m,m-1} = \frac{\partial F_m}{\partial u_{m-1}} = -d(t^{n+1}). \tag{19}$$

Newton's iterative method is then applied to obtain the solution u_j at each time step.

Next, we rewrite the reaction–diffusion equation as a system of two equations by introducing another variable $v = f(u, t)$, that is,

$$F_1(u, v, t) = u_t - v - d(t)u_{xx}, \tag{20}$$

$$F_2(u, v, t) = v - f(u, t). \tag{21}$$

We obtain the fully discrete equations with similar settings as in the single unknown case:

$$F_{1,1}(\mathbf{u}, \mathbf{v}) = \frac{u_1^{n+1} - u_1^n}{\Delta t} - d(t_{n+1})(\frac{u_2^{n+1} - u_1^{n+1}}{(\Delta x)^2}) - v_1^{n+1} = 0, \tag{22}$$

$$F_{1,i}(\mathbf{u}, \mathbf{v}) = \frac{u_i^{n+1} - u_i^n}{\Delta t} - d(t_{n+1})(\frac{u_{i+1}^{n+1} - 2u_i^{n+1} + u_{i-1}^{n+1}}{(\Delta x)^2}) - v_i^{n+1} = 0 \tag{23}$$

where $2 \leq i \leq m - 1$,

$$F_{1,m}(\mathbf{u}, \mathbf{v}) = \frac{u_m^{n+1} - u_m^n}{\Delta t} - d(t_{n+1})(\frac{u_m^{n+1} - u_{m-1}^{n+1}}{(\Delta x)^2}) - v_m^{n+1} = 0, \tag{24}$$

$$F_{2,j}(\mathbf{u}, \mathbf{v}) = f(t_{n+1}, u_j^{n+1} - v_j^{n+1}) = 0, \text{ where } 1 \leq j \leq m. \tag{25}$$

For the discrete system given in Section 2, we can look into the Jacobian blocks. From the definitions given in Section 2.1, the partial derivatives are given for $1 < i < m$ by

$$\frac{\partial F_{1,i}}{\partial u_i} = \frac{1}{\Delta t} + \frac{2d(t_{n+1})}{(\Delta x)^2}, \quad \frac{\partial F_{1,i}}{\partial u_{i-1}} = \frac{\partial F_{1,i}}{\partial u_{i+1}} = -\frac{d(t_{n+1})}{(\Delta x)^2}, \quad \frac{\partial F_{1,i}}{\partial u_j} = 0 \quad \forall j \neq i, i-1.$$

(26)

When $i = 1$,

$$\frac{\partial F_{1,1}}{\partial u_1} = \frac{1}{\Delta t} + \frac{d(t_{n+1})}{(\Delta x)^2}, \quad \frac{\partial F_{1,1}}{\partial u_2} = -\frac{d(t_{n+1})}{(\Delta x)^2}, \quad \frac{\partial F_{1,1}}{\partial u_j} = 0 \quad \forall 3 \leq j \leq m. \quad (27)$$

When $i = m$,

$$\frac{\partial F_{1,m}}{\partial u_m} = \frac{1}{\Delta t} - \frac{d(t_{n+1})}{(\Delta x)^2}, \quad \frac{\partial F_{1,m}}{\partial u_{m-1}} = \frac{d(t_{n+1})}{(\Delta x)^2}, \quad \frac{\partial F_{1,m}}{\partial u_j} = 0 \quad \forall 1 \leq j \leq m-2.$$

(28)

For $1 \leq i \leq m$

$$\frac{\partial F_{1,i}}{\partial v_i} = -1, \quad \frac{\partial F_{1,i}}{\partial v_j} = 0, \quad \forall i \neq j, \tag{29}$$

and

$$\frac{\partial F_{2,i}}{\partial v_i} = -1, \quad \frac{\partial F_{2,i}}{\partial v_j} = 0, \quad \forall i \neq j, \tag{30}$$

and

$$\frac{\partial F_{2,i}}{\partial u_i} = \frac{\partial f}{\partial u_i}, \quad \frac{\partial F_{2,j}}{\partial u_j} = 0, \quad \forall i \neq j. \tag{31}$$

Consequently, J_{12} and J_{22} are negative identity matrices. J_{11} is a symmetric tridiagonal matrix. J_{21} is a diagonal matrix with nonzero elements from the partial derivatives of f with respect to u.

For the split solve, we implement a commonly used splitting scheme to solve a slightly modified problem:

$$\frac{\partial u}{\partial t} - d(t)\frac{\partial^2 u}{\partial x^2} = 0 \tag{32}$$

$$\frac{\partial u}{\partial t} = f(u, t). \tag{33}$$

We again discretize in space with a finite difference method and backward Euler in time, but this time we introduce a dummy variable v into the time discretization to split the two equations in (32)-(33). We obtain the following split discrete form

$$\frac{v_j - u_j^n}{\Delta t} - d(t^{n+1})\frac{v_{j+1} - 2v_j + v_{j-1}}{\Delta x^2} = 0 \tag{34}$$

$$\frac{u_j^{n+1} - v_j}{\Delta t} - f(u_j^{n+1}, t^{n+1}) = 0 \tag{35}$$

so that at the nth time step, we first solve (34) for v and then use it in (35) to solve for u^{n+1}. Note that (34) is linear so only (35) requires a nonlinear solver. A standard Newton iteration is utilized for this nonlinear solve. This is the decoupled solution whose difference with the fully coupled solution is shown in plots below.

In this paper, we test our metrics on a series of specific problems comprising a subset of the following options. The spatial and temporal domains are $[x_0, x_1] = [0, 2\pi]$ and $[t_0, t_1] = [0, 100]$ or $[1, 100]$ if we use t raised to a negative power. We set $a = 0, b = 0$, and $u_0(x) = \sin x$. We choose the forcing function as $f(u, t) = cu^p t^r$, $c \in [10^{-9}, 10^{-6}, 10^{-3}, 1, 10^3, 10^6, 10^9]$, $p \in [-6, -3, 0, 1, 3, 6]$, $r \in [-6, -3, 0, 1, 3, 6]$, and test functions for d, $d(t) = 1$, $d(t) = t$, $d(t) = 2e^{-t}$, or, equivalently, $d(t) = kt^n e^{mt}$ with $(k, n, m) \in [(1, 0, 0), (1, 1, 0), (2, 0, -1)]$.

The spatial and temporal domains are divided into 62 and 1000 subintervals respectively, i.e., $\Delta x \approx 0.1$ and $\Delta t = 0.1$. The ith spatial point is $x = i\Delta x$, and the jth time step is $t = j\Delta t$. Figures 1–11 show the solutions and the log of spatially normalized differences between the coupled solutions $(U(x, t))$ and decoupled solutions $(u(x, t))$ for a set of test cases on 1000 discrete time steps. For fixed time $t = t_k$, the difference $diff_k$ of the solutions from two schemes is calculated using the ℓ2-norm as below:

$$diff_k = \frac{1}{2\pi\sqrt{63}}\sqrt{\sum_{i=1}^{63}(U(x_i, t_k) - u(x_i, t_k))^2}. \tag{36}$$

As this equation takes into account the difference between the coupled and decoupled solutions at every spatial discretization point for each t, it is sufficient to capture the overall behavior of the problem and the dynamic difference between the two solutions.

We consider both linear and nonlinear cases for the reaction function, f. Cases (a)–(e) consider reaction functions that are linear in terms of the solution u and are shown in Figures 1–5. To provide greater complexity in our tests, we also consider cases in which the reaction function is a nonlinear function of the solution. These cases are shown in Figures 6–11. The difference between the coupled and decoupled solutions gives an indication of the coupling strength change over time. The details of the cases are listed below.

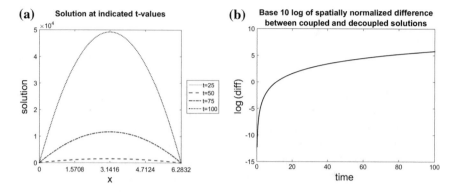

Fig. 1 Solution and log of spatially normalized difference between coupled and decoupled solutions for case (a): $d(t) = 1$ and $f(u, t) = 10^{-6}u^0t^6$. (a) Solution at particular t-values for case (a). (b) Log of spatially normalized difference for case (a).

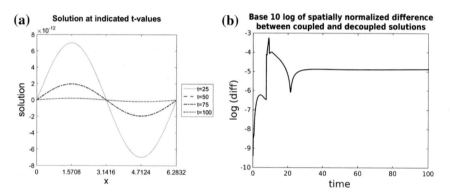

Fig. 2 Solution and log of spatially normalized difference between coupled and decoupled solutions for case (b): $d(t) = 1$ and $f(u, t) = 10^{-6}u^1t^3$. (a) Solution at particular t-values for case (b). (b) Log of spatially normalized difference for case (b).

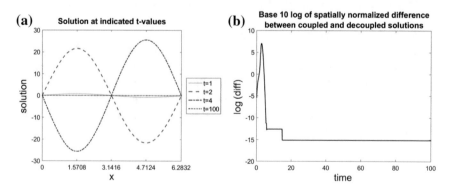

Fig. 3 Solution and log of spatially normalized difference between coupled and decoupled solutions for case (c): $d(t) = t$ and $f(u, t) = 10^0u^1t^3$. (a) Solution at particular t-values for case (c). (b) Log of spatially normalized difference for case (c).

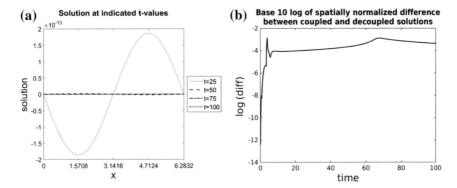

Fig. 4 Solution and log of spatially normalized difference between coupled and decoupled solutions for case (d): $d(t) = t$ and $f(u, t) = 10^{-6}u^1t^6$. (a) Solution at particular t-values for case (d)Solution at particular t-values for case (d). (b) Log of spatially normalized difference for case (d).

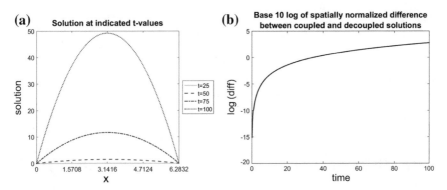

Fig. 5 Solution and log of spatially normalized difference between coupled and decoupled solutions for case (e): $d(t) = 2\exp(-t)$ and $f(u, t) = 10^{-9}u^0t^6$. (a) Solution at particular t-values for case (e). (b) Log of spatially normalized difference for case (e).

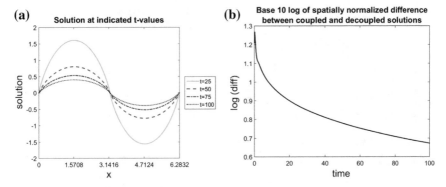

Fig. 6 Solution and log of spatially normalized difference between coupled and decoupled solutions for case (f): $d(t) = 1$ and $f(u, t) = 10^3u^{-1}t^{-1}$. (a) Solution at particular t-values for case (f). (b) Log of spatially normalized difference for case (f).

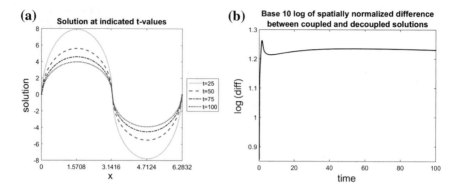

Fig. 7 Solution and log of spatially normalized difference between coupled and decoupled solutions for case (g): $d(t) = 1$ and $f(u, t) = 10^6 u^{-3} t^{-1}$. (a) Solution at particular t-values for case (g). (b) Log of spatially normalized difference for case (g).

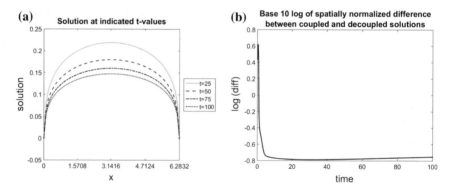

Fig. 8 Solution and log of spatially normalized difference between coupled and decoupled solutions for case (h): $d(t) = t$ and $f(u, t) = 10^{-3} u^{-6} t^{-1}$. (a) Solution at particular t-values for case (h). (b) Log of spatially normalized difference for case (h).

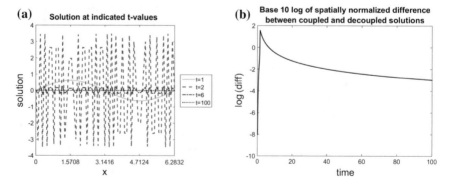

Fig. 9 Solution and log of spatially normalized difference between coupled and decoupled solutions for case (i): $d(t) = 2 \exp(-t)$ and $f(u, t) = 10^0 u^3 t^6$. (a) Solution at particular t-values for case (i). (b) Log of spatially normalized difference for case (i).

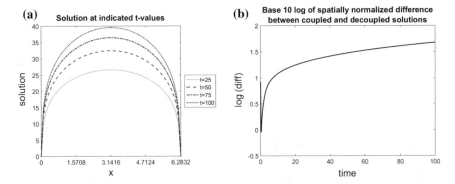

Fig. 10 Solution and log of spatially normalized difference between coupled and decoupled solutions for case (j): $d(t) = 2\exp(-t)$ and $f(u,t) = 10^6 u^{-6} t^3$. (a) Solution at particular t-values for case (j). (b) Log of spatially normalized difference for case (j).

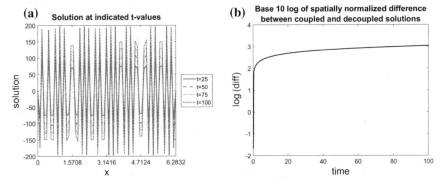

Fig. 11 Solution and log of spatially normalized difference between coupled and decoupled solutions for case (k): $d(t) = 2\exp(-t)$ and $f(u,t) = 10^0 u^3 t^0$.

(a) $d(t) = 1, p = 0, r = 6, c = 10^{-6}$,
(b) $d(t) = 1, p = 1, r = 3, c = 10^{-6}$,
(c) $d(t) = t, p = 1, r = 3, c = 10^0$,
(d) $d(t) = t, p = 1, r = 6, c = 10^{-6}$,
(e) $d(t) = 2\exp(-t), p = 0, r = 6, c = 10^{-9}$,
(f) $d(t) = 1, p = -1, r = -1, c = 10^3$,
(g) $d(t) = 1, p = -3, r = -1, c = 10^6$,
(h) $d(t) = t, p = -6, r = -1, c = 10^{-3}$,
(i) $d(t) = 2\exp(-t), p = 3, r = 6, c = 10^0$,
(j) $d(t) = 2\exp(-t), p = -6, r = 3, c = 10^6$, and
(k) $d(t) = 2\exp(-t), p = 3, r = 0, c = 10^0$.

In case (a), we observe that the difference between the solutions is increasing over time, which shows that the coupling strength is getting stronger. Case (b) indicates that coupling is becoming stronger when $t \leq 15$ and remains relatively unchanged after $t = 25$. In case (b), the peak of log of the difference is 10^{-3}, which means the difference is sufficiently small. So, the decoupled method is fine to replace the coupled method. In case (c), we note that between $t = 1$ and $t = 2$, a dramatic gap appears, which means that after $t = 2$, the processes are tightly coupled. However, for later times, the difference in solutions is smaller, indicating that the decoupled method would be sufficient. In case (d), the error increases sharply before $t \leq 8$ and slowly afterward. The decoupled method is acceptable for all time. Case (e) indicates that the physical processes are more tightly coupled for $t \geq 15$, but generally, the decoupled method could be used to substitute since the maximum difference is less than 10^{-5}.

The solutions and log of the difference between coupled and decoupled solutions for the nonlinear cases are shown in Figures 6–11, respectively. In case (f), as shown in Figure 6, the difference decreases gradually over time which indicates that the coupling strength decreases with time. As shown in Figure 7(b), the difference in solutions in case (g) increases drastically for small t-values, then decreases quickly before increasing slightly, leveling out, and slowly decreasing for the rest of time; this behavior shows that the coupling increases dramatically initially and then mostly decreases at varying speeds as time increases.

In case (h), as shown in Figure 8, the difference in solutions decreases quickly for a short period of time before increasing slowly as time increases; this indicates that the coupling decreases initially then increases slowly for the rest of time. In Figure 9(b), we see for case (i) that although the difference increases for small time, it decreases gradually as time increases which would indicate that the coupling decreases with time after increases for a short period when t is small.

In case (j), shown in Figure 10, the difference decreases quickly for small t and increases for the rest of time; this shows that the coupling decreases for small t before increasing steadily. In the last case, as shown in Figure 11, the difference increases for all time which indicates the coupling increases as time increases.

We have shown the solution profiles and the difference between coupled and decoupled schemes from numerical experiments with both linear and nonlinear cases for the reaction function. The plots of log of difference interest us the most. From these plots, it is easily observed how the coupling strength changes over time. Depending on the test case, the decoupled scheme is applicable either for all time or for only some certain ranges of time steps. In other words, the coupling strength highly depends on the parameter scales and the form of $d(t)$. The plots are important to determine the effectiveness of each of the metrics and will be explained with more details in the following sections.

4 Numerical Results

In this section, we detail numerical results on our test cases using the metrics presented in Section 2.

4.1   Off-Diagonal Jacobian Blocks

From the partial derivatives of the discrete system for the reaction–diffusion problem given in (26)–(31), we have

$$
J_{12} =
\begin{bmatrix}
-1 & 0 & \cdots & 0 \\
0 & -1 & \cdots & 0 \\
0 & \cdots & \ddots & 0 \\
0 & \cdots & 0 & -1
\end{bmatrix}
= -\mathbf{I}_n
\tag{37}
$$

and

$$
J_{21} =
\begin{bmatrix}
\left.\frac{\partial f}{\partial u_1}\right|_{t_n,\mathbf{u}^n} & 0 & \cdots & 0 \\
0 & \left.\frac{\partial f}{\partial u_2}\right|_{t_n,\mathbf{u}^n} & \cdots & 0 \\
0 & \cdots & \ddots & 0 \\
0 & \cdots & 0 & \left.\frac{\partial f}{\partial u_m}\right|_{t_n,\mathbf{u}^n}
\end{bmatrix}
\tag{38}
$$

which seem to indicate that the diffusion physics is coupled to the reaction physics constantly in time since $\|J_{12}\| = 1$, but the coupling of the reaction physics to the diffusion physics can be weaker or stronger (changing in time) depending on $\|J_{21}\|$. Any matrix norm can be used to compute $\|J_{21}\|$ since any norm would show the same overall behavior. For the results below, the 2-norm of J_{21} is used to compute the values shown in the plots. Notice that since J_{21} is a diagonal matrix, the 2-norm of J_{21} is equivalent to taking the largest element in the matrix. Since $\|J_{12}\| = 1$ for all cases of reaction and diffusion functions, we present only the results for $\|J_{21}\|$ below.

4.1.1   Linear Test Cases

The linear cases of f produce two types of results for $\|J_{21}\|$: either $\|J_{21}\| = 0$ for all time or $\|J_{21}\|$ depends on t but not on u.

For case (a) when $d(t) = 1$ and $f(u,t) = 10^{-6}t^6$, the reaction function is not dependent on u. Consequently, $\frac{\partial f}{\partial u_i} = 0$ for all i so that $\|J_{21}\| = 0$ for all time (figure omitted). Using this $\|J_{21}\|$ as a strength of coupling metric implies that there is no coupling of the reaction physics to the diffusion physics. As we always have $\|J_{12}\| = 1$, then there is at least a constant coupling of the diffusion physics to the reaction physics indicated by this metric. However, in this case, the metric does not suggest an increase in coupling with time as is indicated by the log difference in Figure 1. If we take into account the solution behavior in this case shown in Figure

Fig. 12 Norm of J_{21} block
over time for case (b):
$d(t) = 1$ and
$f(u, t) = 10^{-6} u t^3$.

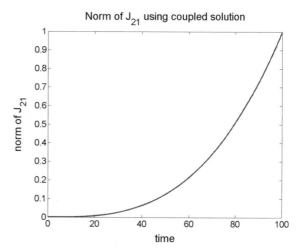

1(a) by considering a relative difference, the relative difference would still increase with time which is not matched by the off-diagonal Jacobian blocks metric.

In case (b) when $d(t) = 1$ and $f(u, t) = 10^{-6} u t^3$, we have $\frac{\partial f}{\partial u_i} = 10^{-6} t^3$ for all i. Then, $\|J_{21}\|$ varies with time as given in Figure 12. In this case, $\|J_{21}\|$ is not dependent on u so the results are the same whether the fully coupled or weakly coupled solution method is used. The behavior shown in Figure 12 would imply an increase in the coupling over time which is contrary to the constant coupling following initial changes indicated by the log difference plot in Figure 2. Considering the solution behavior shown in Figure 2(a) along with the log difference behavior, the relative difference would be larger than the log difference shown in Figure 2 since the solution approaches a zero equilibrium. Although the relative difference would still be constant long term, the larger magnitude matches up somewhat better with the metric results; however, the metric still implies continually increasing coupling which does not match up with the log difference or relative difference indications.

For case (c) when $d(t) = t$ and $f(u, t) = u t^3$, then $\frac{\partial f}{\partial u_i} = t^3$ for all i. Again, the reaction function has changed from the previous two cases by only a constant factor. The diffusion function d has also changed in this case which results in different solutions. However since the diffusion function does not affect this metric, the metric behavior in this case is the same as the previous two cases with the only difference being the magnitude of the values. Consequently, the plot of $\|J_{21}\|$ is again omitted. The results shown in Figure 3 indicate a decrease in coupling for large t which is contrary to the metric results in this case. If we also consider the solution behavior shown in Figure 3(a), the relative difference would increase in magnitude, but still have a large increase for small t followed by a decrease over time. Consequently, the off-diagonal Jacobian blocks metric does not provide good predictions in this case.

With case (d), $d(t) = t$ and $f(u, t) = 10^{-6} u t^6$ so $\frac{\partial f}{\partial u_i} = 10^{-6} t^6$ for all i, so $\|J_{21}\|$ depends on time (see Figure 13). $\|J_{21}\|$ does not depend on u so the results are the

Fig. 13 Norm of J_{21} block over time for case (d): $d(t) = t$ and $f(u, t) = 10^{-6} u t^6$.

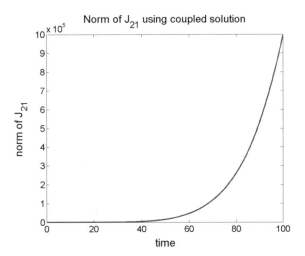

same whether the fully coupled or weakly coupled solution method is used. The values for $\|J_{21}\|$ seem to indicate an increase in coupling over time. Although this indication matches up somewhat with initial coupling changes implied by Figure 4, the long-term decrease in coupling is not predicted by this metric. If we take into consideration the solution behavior shown in Figure 4(a), the relative difference would indicate somewhat increased coupling overall since the solution approaches equilibrium at zero. Nonetheless, the long-term behavior would still indicate a slow decrease in coupling which is not indicated by the off-diagonal Newton blocks metric in this case.

For the last linear case (e), when $d(t) = 2e^{-t}$ and $f(u, t) = 10^{-9} t^6$, the reaction function is not dependent on u, so the results are the same as case (a). Consequently, figures are omitted. Just as in that case, the only coupling indicated by this metric is the constant coupling of the diffusion physics to the reaction physics. This constant coupling does match up with the long-term coupling behavior shown in Figure 5, but it does not capture any of the initial increases in coupling. If we take into consideration the solution behavior shown in Figure 5(a), the relative difference would increase with time nearly linearly because of the increase in the solution values. Consequently, if we consider the relative difference as an indication of the coupling, then the off-diagonal Jacobian metric does not produce good predictions in this case.

4.1.2 Nonlinear Test Cases

The nonlinear cases for f produce remarkably different results for $\|J_{21}\|$ from the linear cases since $\frac{\partial f}{\partial u}$ is now a function of the solution u. Because of this dependence on u, these cases produce different results depending on which solution is used, that is, the solution obtained using the fully coupled discretization scheme or the solution obtained using the split scheme. Therefore, results obtained using both the coupled solution and the split solution are provided for each case.

This dependence on u also produces some instabilities in the metric computation. Since many of the nonlinear cases have u raised to a negative power, instabilities in the metric computation arise when u is equal to or close to 0. To resolve this issue, $\frac{\partial f}{\partial u}$ is computed at each time step only at the spatial values for which u was larger than some given tolerance. In the results shown below, we use a value of 10^{-3} for the tolerance to resolve nearly all of the instability arising from a solution close to zero while still preserving the overall behavior of the metric. Occasionally, some instability still arises when using the solution obtained from the split scheme; however, this only appears for small t values when t is also raised to a negative power. In these cases, in order to see the long-term behavior of the metric, the values of $\|J_{21}\|$ are plotted omitting the first few time steps; the specific number of time steps omitted varies depending on the case, but the maximum number of time steps omitted for all the test cases is 17.

For case (f) when $d(t) = 1$ and $f(u, t) = 10^3 u^{-1} t^{-1}$, $\frac{\partial f}{\partial u} = -10^3 u^{-2} t^{-1}$ so $\|J_{21}\|$ depends on both the solution u and t. The values of $\|J_{21}\|$ over time are given in Figure 14(a) contains the results obtained using the coupled solution, while Figure 14(b) uses the split solution. The metric values obtained using the coupled solution appear to increase linearly in time. However, the values obtained from the split solution increase quickly at first until essentially plateauing at a magnitude of 10. Although the rates of increase are different, the values of $\|J_{21}\|$ from both solutions indicate an increase in coupling over time, yet the log difference values in Figure 6(b) indicate the opposite behavior with a decrease in coupling. If we also consider the solution behavior shown in Figure 6(a), the solution decreases to equilibrium at zero faster than the log difference decreases which results in a very slow increase in the relative difference. If we consider the relative difference as an indication of the coupling, then the off-diagonal Jacobian blocks metric produces better predictions in this case.

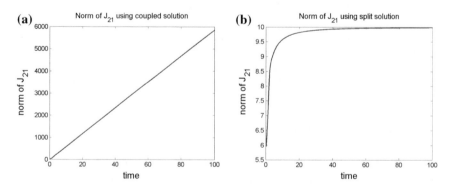

Fig. 14 Norm of J_{21} block over time for case (f): $d(t) = 1$ and $f(u, t) = 10^3 u^{-1} t^{-1}$. (a) Using solution result from coupled scheme. (b) Using solution result from split scheme.

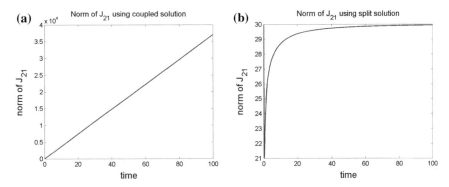

Fig. 15 Norm of J_{21} block over time for case (g): $d(t) = 1$ and $f(u, t) = 10^6 u^{-3} t^{-1}$ (a) Using solution result from coupled scheme. (b) Using solution result from split scheme.

With case (g) when $d(t) = 1$ and $f(u, t) = 10^6 u^{-3} t^{-1}$, then $\frac{\partial f}{\partial u} = -3 \cdot 10^6 u^{-4} t^{-1}$. The values of $\| J_{21} \|$ using both the coupled and split solution are shown in Figure 15. The same overall behavior as the previous case is shown for both the results using the coupled solution in Figure 15(a) and the results using the split solution in Figure 15(b), although the magnitudes are larger in both. Again, these metric results suggest an increase in coupling with time. In this case, these results match up fairly well for small t with the coupling indicated by Figure 7(b), but the long-term slow decrease in coupling is not shown by these metric results. If we consider the relative difference instead by incorporating the solution behavior shown in Figure 7(a), then increased coupling over time is indicated since the solution decreases to 0 over time resulting in an increasing relative difference. Again, this metric produces more accurate predictions if the solution behavior is incorporated into the evaluation.

With case (h) when $d(t) = t$ and $f(u, t) = 10^{-3} u^{-6} t^{-1}$, we see that $\frac{\partial f}{\partial u} = -6 \cdot 10^{-3} u^{-7} t^{-1}$ which gives values of $\| J_{21} \|$ shown in Figure 16. Again, we see the same overall behavior as the previous two cases with larger magnitudes than the first cases. Although using the coupled solution (Figure 16(a)) gives the same magnitudes as the coupled solution in the previous case (Figure 15(a)), the split solution (Figure 16(b)) gives even large magnitudes than previous cases, and the plateau is more gradual. The long-term increase in coupling shown in Figure 8(b) correlates fairly well with the increase in coupling indicated by these metric results in this case; however, none of the details, such as the initial decrease in coupling or the very slow increase, are captured by the values of $\| J_{21} \|$ here. Even if we take into account the solution behavior shown in Figure 8(a), the relative difference behaves similarly to the log difference plot except the magnitudes over all are larger and the increase over time is faster since the solution decreases to zero. Consequently, the metric results still do not make the correct predictions.

When we consider case (i) with $d(t) = 2 \exp(-t)$ and $f(u, t) = u^3 t^6$, we obtain $\frac{\partial f}{\partial u} = 3u^2 t^6$ which produces results shown in Figure 28. Due to oscillations in the solution obtained from the split scheme for small t, the values of $\| J_{21} \|$ obtained using

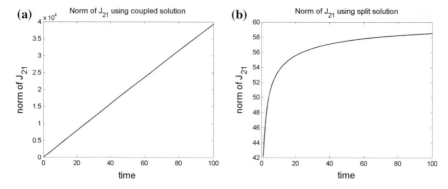

Fig. 16 Norm of J_{21} block over time for case (h): $d(t) = t$ and $f(u, t) = 10^{-3}u^{-6}t^{-1}$ (a) Using solution result from coupled scheme. (b) Using solution result from split scheme.

the split solution have large erratic spikes for small t before becoming identically zero just before $t = 20$. Although this behavior is caused by oscillations in the split solution, it matches up better with the long-term coupling behavior indicated by the log difference plot for this case in Figure 9(b). If we incorporate the solution behavior shown in Figure 9(a), the decrease in the solution is approximately the same rate as the decrease in the difference resulting in the relative difference being constant for large time after a large spike at first. Again, this matches up better with the behavior indicated by the metric results from the split solution but does not coincide at all with the results from the coupled solution.

With case (j) when $d(t) = 2\exp(-t)$ and $f(u, t) = 10^6 u^{-6}t^3$, we see the usual linear increase in $\|J_{21}\|$ using the coupled solution (Figure 18(a)) and a return to the usual increase to a plateau in $\|J_{21}\|$ using the split solution (Figure 18(b)). The metric results for this case correspond to the coupling behavior indicated by the log difference plot in Figure 10(b) better than in any other case, especially when one considers that the $\|J_{21}\|$ values were computed using the split solution. Considering the solution behavior shown in Figure 10(a), the relative difference behavior is similar to the log difference behavior although with smaller magnitudes because of the increase in the solution. Consequently, the off-diagonal Jacobian blocks metric produces good predictions when considering the relative difference as well (Figure 17).

The last nonlinear case (k), when $d(t) = 2\exp(-t)$ and $f(u, t) = u^3$, we have $\frac{\partial f}{\partial u} = 3u^2$. The behavior for $\|J_{21}\|$ using the coupled solution is the same as all the other nonlinear cases, but the behavior using the split solution is drastically different. The splitting scheme in this case causes the split solution to be drastically different from the coupled solution which is evidenced by the log difference behavior and magnitude in Figure 11(b). This drastic difference in turn causes the difference in the two $\|J_{21}\|$ behaviors shown in Figure 19. In this case, the $\|J_{21}\|$ values exhibited in Figure 19(a) better match the coupling strength evidenced by the log difference behavior than the $\|J_{21}\|$ values shown in Figure 19(b). If we consider the solution behavior shown in Figure 11(a), the relative difference is large initially since the

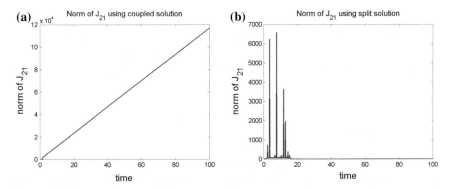

Fig. 17 Norm of J_{21} block over time for case (i): $d(t) = 2\exp(-t)$ and $f(u, t) = u^3 t^6$. (a) Using solution result from coupled scheme. (b) Using solution result from split scheme.

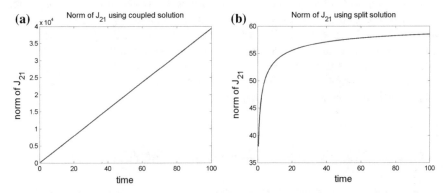

Fig. 18 Norm of J_{21} block over time for case (j): $d(t) = 2\exp(-t)$ and $f(u, t) = 10^6 u^{-6} t^3$. (a) Using solution result from coupled scheme. (b) Using solution result from split scheme.

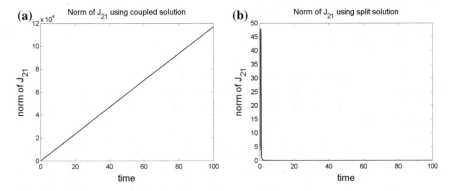

Fig. 19 Norm of J_{21} block over time for case (k): $d(t) = 2\exp(-t)$ and $f(u, t) = u^3$. (a) Using solution result from coupled scheme. (b) Using solution result from split scheme.

solution starts near zero and then decreases slowly as time increases because the magnitude of the solution increases as time increases. Consequently, the off-diagonal Jacobian metric results obtained using the split solution matches up better if we incorporate the solution behavior into the comparison, although neither metric results match up perfectly when comparing to a relative difference.

4.1.3   Summary

In nearly all the cases, linear and nonlinear, the $\|J_{21}\|$ metric continuously increases over time although the rates of the increases change somewhat depending on the case or whether the split or coupled solution is considered. Sometimes, the behavior exhibited by this metric matches up with the coupling strengths indicated by corresponding log difference plots, but other times, the behavior is nearly opposite. Generally, the metric predictions match up better with coupling behavior indicated by a relative difference obtained by incorporating solution behavior. One case shows a good match between the metric results and both the log difference and relative difference results. More often than not, this metric gives poor results for predicting coupling in the problems considered although the results are somewhat better if a relative difference is used for comparison.

4.2   Condition Number of Diagonal Jacobian Blocks

Next, we test the metric discussed in Section 2.2. For the reaction diffusion problem, we mainly focus on the condition number J_{11} as described in (3). The results shown are computed using the decoupled solution.

4.2.1   Linear Test Cases

From Section 3, we observe that when $d(t) = 1$, the condition number of J_{11} will be constant. For cases (a)–(b), this metric is constant (figures omitted) and does not provide sufficient information to determine the coupling strength nor the change of the coupling strength.

When $d(t) = t$, in case (c), this metric does provide some evidence that the coupling is stronger over time before $t = 2$. The condition number of J_{11} reaches $10^{1.8}$, which indicates the split method is acceptable. However, it fails to capture the sharp change around $t = 2$. From Figure 3, we see that when $t = 2.5$, the difference between the two solutions is larger than 10^5, but this information is not observed in Figure 20. In case (d), this metric shows the trend of coupling strength change between $t = 0$ and $t = 10$. Since the condition number increases quickly, this metric corresponds nicely with the observations from Figure 4. However, since the condition number is approaching 10^3 after $t \geq 40$, the metric indicates the decoupled

Fig. 20 Condition number of J_{11} for case (c): $d(t) = t$ and $f(u, t) = 10^0 u^1 t^3$.

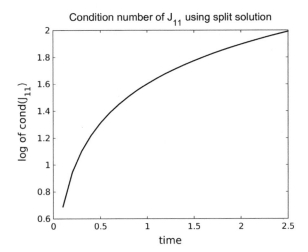

method is not working in this case. This conclusion is opposite from what we see in Figure 4(b), where the split method is acceptable for all time. For case (e), the condition number is decreasing over time, which is opposite to the trend of the solution difference change as shown in Figure 5, but due to the relatively small value of condition number of J_{11}, the split method is still considered to work which also corresponds to the conclusions of Figure 5.

For the linear tests, we conclude that the condition number of J_{11} successfully predicts the coupling strength over time in case (e). For case (c), the results from this metric show that it fails to predict the sharp change of coupling strength. The result for case (d) predicts the opposite conclusion to the true solution. Last, when $d(t)$ is the constant 1, the condition number is a constant which cannot provide enough information to determine any change of the coupling strength, as in cases (a)–(b) (Figures 21 and 22).

4.2.2 Nonlinear Test Cases

We observe that when $d(t) = 1$, the condition number of J_{11} remains unchanged as seen in the linear test cases (figures for cases (f) and (g) omitted). Again, the condition number cannot help to determine the coupling strength. When $d(t) = t$ for test case (h), we see from Figure 8 that the solution differences decrease sharply at first and then level off. The results of the metric when $t < 10$ coincide with these facts as shown in Figure 23 where the maximum condition number of J_{11} also increases sharply at first and then levels off. But it fails to indicate the correct choice when $t > 10$. In case (i), this metric shows the trend of coupling strength decreasing as time increases which matches the result from Figure 9 after $t \geq 2$, but this metric again fails to predict the sharp change before $t = 2$. In cases (j) and (k) shown in Figures 25 and 26, the condition numbers fall sharply before $t = 5$ and remain unchanged

Fig. 21 Condition number
of J_{11} for case (d): $d(t) = t$
and $f(u, t) = 10^{-6}u^1t^6$.

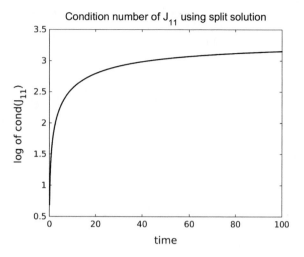

Fig. 22 Condition number
of J_{11} for case (e):
$d(t) = 2e^{-t}$ and
$f(u, t) = 10^{-9}u^0t^6$.

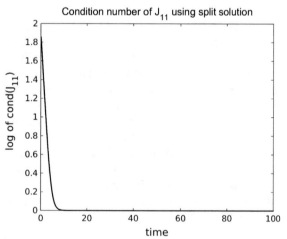

to around 1 afterward which may suggest a weaker coupling for $t \geq 5$. In these
two cases, the condition number suggests the use of the split method after $t = 5$.
However, Figures 10–11 show that the coupling is getting stronger for $t \leq 5$, and the
solution differences are more than $10^{1.5}$ for almost all times. The metric gives the
incorrect indication of the choice of decoupling method as well as the opposite trend
of coupling strength change (Figure 24).

4.2.3 Summary

From the above tests, we find that the condition number of J_{11} has successfully
predicted the acceptance of a decoupled solution in case (e). For test cases (c), (d),

Fig. 23 Condition number of J_{11} for case (h): $d(t) = t$ and $f(u, t) = 10^{-3}u^{-6}t^{-1}$.

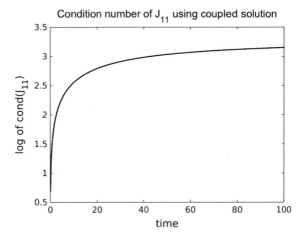

Fig. 24 Condition number of J_{11} for case (i): $d(t) = 2e^{-t}$ and $f(u, t) = 10^0 u^3 t^6$.

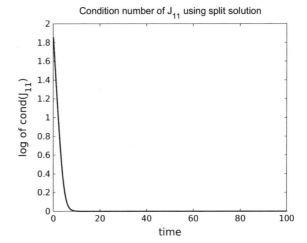

and (h), it gives partially correct indications for the beginning part and fails to predict the change of coupling strength for later times. For test case (i), it provides correct indication for later time steps. For test cases (j) and (k), this metric provides a totally opposite prediction to the feasibility of split methods as well as the coupling strength change. When $d(t) = 1$ as in cases (a)–(b) and (f)–(g), the condition numbers are constant; thus, this metric cannot provide sufficient information to understand the coupled processes.

As a summary, this metric works fine to suggest the feasibility of split method only in some certain test cases. It may predict the correct choice for part of the simulation. There are also cases where this metric gives opposite conclusion. For example, when $d(t)$ is a constant, the condition number stays the same and fails to provide any information. More often than not, this metric does not give good results to determine whether to take split scheme or not with consideration to the total simulation time.

Fig. 25 Condition number of J_{11} for case (j): $d(t) = 2e^{-t}$ and $f(u, t) = 10^6 u^{-6} t^3$.

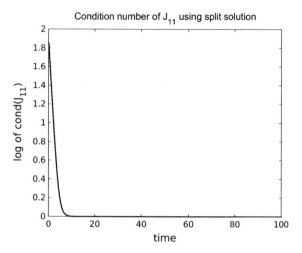

Fig. 26 Condition number of J_{11} for case (k): $d(t) = 2e^{-t}$ and $f(u, t) = 10^0 u^3 t^0$.

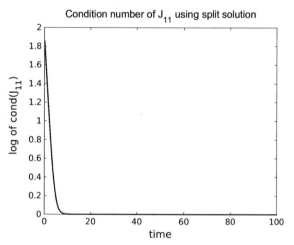

On the other hand, as we can see in the log of difference plots, there may be sharp changes which will never be seen in the curves of condition number. This means that the condition number metric is unable to indicate the coupling strength change either.

4.3 Time scales

The timescales for the diffusion and reaction operators in the diffusion–reaction equation can be computed using a simple dimensional argument as in [4, 6]. Similar to [4], we will again define the characteristic timescales as the absolute values of the operator timescales. Then, we have the following characteristic timescales for the reaction–diffusion equation given in (7):

$$\text{Characteristic Diffusion Timescale:} \tau_d = \left| \frac{d(t)u_{xx}}{u} \right|^{-1} = \left| \frac{u}{d(t)u_{xx}} \right|,$$

$$\text{Characteristic Reaction Timescale:} \tau_f = \left| \frac{f}{u} \right|^{-1} = \left| \frac{u}{f} \right|,$$

$$\text{Characteristic Dynamic Timescale:} \tau_{dyn} = \left| \frac{1}{u} \frac{\partial u}{\partial t} \right|^{-1}.$$

According to Ropp, Shadid, and Ober [6], if the reaction timescale is orders of magnitude smaller than the diffusion time scale (i.e., $\tau_f << \tau_d$), then the dynamic timescale is approximately the same as the reaction timescale (i.e., $\tau_{dyn} \approx \tau_f$). On the other hand, if the diffusion timescale is much smaller than the reaction timescale (i.e., $\tau_d << \tau_f$), then the dynamic timescale is now on the order of the diffusion timescale (i.e., $\tau_{dyn} \approx \tau_d$). In other words, if one of the timescales is much smaller than the other, then the dynamic timescale is expected to be on the order of the smaller one.

However in the case when the diffusion and reaction timescales are on the same order (i.e., $\tau_d \approx \tau_f$), then the dynamic timescale can be drastically different, sometimes even much larger than either τ_d or τ_f (see [6] for details of timescale analysis). In essence, the dynamic timescale (and overall end behavior) in this case is highly dependent on the interaction of the diffusion and reaction physics. This may mean that the timescales being on the same magnitude indicates a higher coupling between the two physics.

To compute the diffusion and reaction timescales at a certain time, finite difference approximations are computed for every spatial discretization point. To find an approximate timescale at a certain time step, the finite difference approximations are averaged/normed over all the spatial discretization points. The approximate time scale values vary sometimes depending on the averaging/norm chosen. Consequently, three different approaches are used to compute the time scales at each time step:

- The maximum value given by the finite difference approximation over all spatial discretization points.
- The average value over all spatial discretization points.
- The 2-norm value over all spatial discretization points.

To determine whether the reaction and diffusion timescales are on the same magnitude at a certain time $t = t_n$, the base ten logarithm of the quotient of the two timescales, i.e., $\log\left(\frac{|\tau_d|_n}{|\tau_f|_n}\right)$, is computed where the magnitude notation $|\cdot|$ represents any of the three approaches described in the previous paragraph and the subscript n denotes the value obtained by using the solution computed at $t = t_n$. If this value is "close" to 0 (or even simply less than 1 in magnitude), then the two timescales are "approximately" the same. Since the dynamic timescale depends much on the comparison of the two timescales, the logarithm values may give insight into the coupling of the problem. Plots of $\log\left(\frac{\tau_d}{\tau_f}\right)$ over time for the test cases described in Section 3 are provided below in the order that the cases are listed in Section 3.

Since this metric involves many divisions, there is great potential for instability in the computations if the value in the denominator is ever zero or close to zero. We use two methods to deal with this issue.

The first method simply ignores all the spatial discretization points at which the denominator is less than a certain tolerance when computing a quotient before the magnitude of τ_f and τ_d are obtained; that is, the necessary quotient is computed only at spatial discretization points for which the denominator is greater than a set tolerance. The final quotient $\frac{|\tau_d|_n}{|\tau_f|_n}$, which no longer involves spatial discretization points, sets small values in τ_f (those less than the tolerance) to be equal to the tolerance and computes the quotient at all time values; this last quotient does not simply ignore the values for which the denominator is too small as is done in the previously computed quotients because we wish to obtain a metric value at every time step. We refer to this method as the reduction method.

The second method deals with all quotients the same way: It sets small values in the denominator (those less than a given tolerance) to be equal to the given tolerance value. We refer to this method as the reset method.

The value for the tolerance in each case is chosen to be large enough to allow the metric to be computed at more time steps, but small enough to not change the overall behavior of the metric. Even with these stabilizing techniques, occasional spikes in the values of $\log\left(\frac{\tau_d}{\tau_f}\right)$ are unavoidable without changing the overall behavior of the metric.

4.3.1 Linear Test Problems

The results from the linear test cases fall into three general categories. The first and most common category is an almost asymptotic behavior in which the metric values increase continuously over time. The second category is similar to the first in that the values increase over time, but the values plateau much more than in the first category. The third category increases for all time also, but the increase is basically linear in time.

For case (a) when $d(t) = 1$ and $f(u, t) = 10^{-6}t^6$, no stabilization is needed to produce the results shown in Figure 27. The results using the coupled solution and the split solution are nearly identical except for a few minor instabilities for small t in the metric using the split solution. Both results have a sharp initial increase followed by a near plateau around one although the values do continue increasing very slowly. The results indicate a sharp increase in coupling for a short period of time and then a very slow decrease in coupling as time progresses. This indication matches up fairly well with the log difference results in Figure 1 although that plot

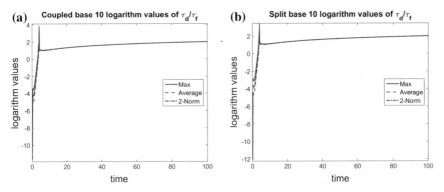

Fig. 27 Plots of $\log\left(\frac{\tau_d}{\tau_f}\right)$ over time for case (a): $d(t) = 1$ and $f(u,t) = 10^{-6}t^6$. (a) Using coupled solution. (b) Using split solution.

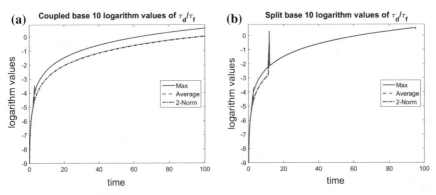

Fig. 28 Plots of $\log\left(\frac{\tau_d}{\tau_f}\right)$ over time for case (b): $d(t) = 1$ and $f(u,t) = 10^{-6}ut^3$. (a) Using coupled solution. (b) Using split solution.

indicates an increase in coupling for all time. If we take into account the solution behavior in this case shown in Figure 1(a) by considering a relative difference, the results match up better since the relative difference would increase slower because the solution increases also as time increases.

When we consider case (b) with $d(t) = 1$ and $f(u,t) = 10^{-6}ut^3$, stabilization is required for both cases shown in Figure 28. We use the reduction method for both the coupled solution and the split solution with the tolerance set to 1e-325. Essentially, this tolerance omits only values identically 0 which allows for the split solution to produce metric results for the largest t interval. Ignoring the instabilities in the results from the split solution, the metric results for both the cases have the same overall behavior, asymptotically approaching a value slightly larger than zero. These results indicate a steady increase in coupling for all time approaching a very strong coupling with larger t; this prediction initially matches up with the coupling implications of the log difference plot in Figure 2 but does not coincide with the

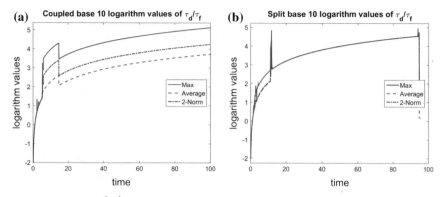

Fig. 29 Plots of $\log\left(\frac{\tau_d}{\tau_f}\right)$ over time for case (c): $d(t) = t$ and $f(u, t) = ut^3$. (a) Using coupled solution. (b) Using split solution.

constant coupling that occurs shortly after $t = 20$. Considering the solution behavior shown in Figure 2(a) along with the log difference behavior, the relative difference would be larger than the log difference shown in Figure 2 since the solution approaches a zero equilibrium. Although the relative difference would still be constant long term, the larger magnitude implies a stronger coupling than is illustrated by the log difference plot. Consequently, the timescales metric results coincide well with the coupling indicated by the difference and solution behaviors.

With case (c) when $d(t) = t$ and $f(u, t) = ut^3$, we again employ the reduction method with a tolerance of 1e-325 to produce metric results using both the coupled solution and the split solution. In this case, there are instabilities still evident in both the coupled solution and split solution results shown in Figure 29. The overall behavior shows an asymptotic increase in both cases to a value around four. Since the metric initially has values with magnitude less than one and increases to values farther away from zero, it indicates a strong initial coupling followed by a decrease in coupling with time. This matches fairly well with the log difference results shown in Figure 3 since the values in that plot initially increase for small t and then decrease with time. If we also consider the solution behavior shown in Figure 3(a), the relative difference would increase in magnitude, but still have a large increase for small t followed by a decrease over time.

For case (d) $d(t) = t$ and $f(u, t) = 10^{-6}ut^6$, stabilization is used for both results shown in Figure 30. The reduction method with a tolerance of 1e-325 is used again. Ignoring the instabilities, the behavior shows an asymptotic increase to a value slightly above four. This behavior suggests an initial increase in coupling that decreases with time. The prediction in this case matches up somewhat with the indication of the log difference plot in Figure 4; however, the long-term coupling prediction does not coincide. If we take into consideration the solution behavior shown in Figure 4(a), the relative difference would indicate somewhat increased coupling overall since the solution approaches equilibrium at zero. Nonetheless, the long-term

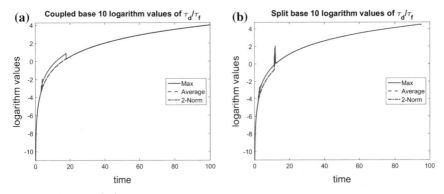

Fig. 30 Plots of $\log\left(\frac{\tau_d}{\tau_f}\right)$ over time for case (d): $d(t) = t$ and $f(u, t) = 10^{-6}ut^6$. (a) Using coupled solution. (b) Using split solution.

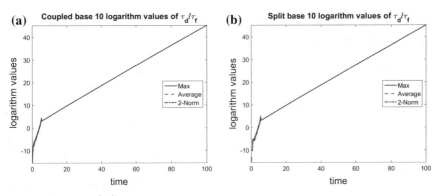

Fig. 31 Plots of $\log\left(\frac{\tau_d}{\tau_f}\right)$ over time for case (e): $d(t) = 2\exp(-t)$ and $f(u, t) = 10^{-9}t^6$. (a) Using coupled solution. (b) Using split solution.

behavior would still indicate a nearly constant coupling which is not fully indicated by the timescales metric in this case.

With case (e) when $d(t) = 2\exp(-t)$ and $f(u, t) = 10^{-9}t^6$, no stabilization is used to produce the plots shown in Figure 31. In this case, the values of $\log\left(\frac{\tau_d}{\tau_f}\right)$ start closer to 0 and increase linearly as t grows. These results indicate a very strong initial coupling that decreases linearly with time. Contrary to this, Figure 5 indicates an initial increase in coupling that levels out to a constant level of coupling long term. Even if we take into consideration the solution behavior shown in Figure 5(a), the relative difference would still increase with time, albeit at a slower rate, reinforcing the indication of increased coupling over time.

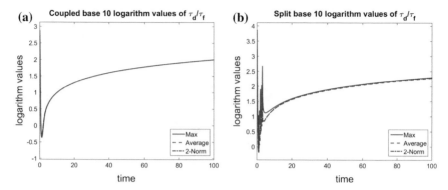

Fig. 32 Plots of $\log\left(\frac{\tau_d}{\tau_f}\right)$ over time for case (f): $d(t) = 1$ and $f(u, t) = 10^3 u^{-1} t^{-1}$. (a) Using solution result from coupled scheme. (b) Using solution result from split scheme.

4.3.2 Nonlinear Test Cases

The nonlinear test cases essentially fall into the same three categories as we see with the linear test cases. Much of the differences, e.g., the sharp initial decrease in coupled solution results and the initial spikes in the split solution results, are most likely due to instabilities in the calculations. The most notable difference is in the fact that very few of the nonlinear cases require stabilization; more specifically, only the two cases which use a positive power of u require stabilization to produce the results.

For case (f) when $d(t) = 1$ and $f(u, t) = 10^3 u^{-1} t^{-1}$, the results are obtained without any stabilization and are shown in Figure 32. Except for the instabilities in the split solution results, the overall behavior is the same in both, starting near zero and increasing asymptotically to a value slightly above two in the case using the coupled solution and around 2.25 in the case using the split solution. This seems to indicate a strong initial coupling that decreases with time which matches up very well with the coupling indicated by the log difference plot in Figure 6(b). However, if we also consider the solution behavior shown in Figure 6(a), the solution decreases to equilibrium at zero faster than the log difference decreases which results in a very slow increase in the relative difference. If we consider the relative difference as an indication of the coupling, then the timescales metric does not produce good predictions in this case.

Figure 33 shows the results for case (g) when $d(t) = 1$ and $f(u, t) = 10^6 u^{-3} t^{-1}$; no stabilization is used in this case. The overall behavior in both plots is similar to the previous case with the asymptotic increase now approaching a value slightly above 2.5 in the case using the split solution. This again indicates a strong initial coupling decreasing as time progresses. However in this case, the indicated coupling does not fully match up with the implications from Figure 7(b) which indicates an initial increase followed by a very slow decrease in coupling. If we consider the relative difference instead by incorporating the solution behavior shown in Figure 7(a),

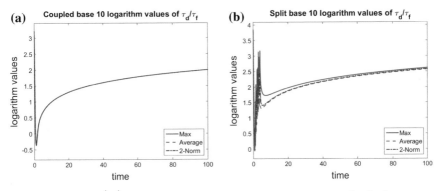

Fig. 33 Plots of $\log\left(\frac{\tau_d}{\tau_f}\right)$ over time for case (g): $d(t) = 1$ and $f(u, t) = 10^6 u^{-3} t^{-1}$. (a) Using solution result from coupled scheme. (b) Using solution result from split scheme.

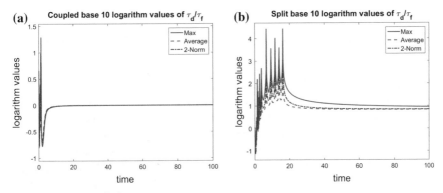

Fig. 34 Plots of $\log\left(\frac{\tau_d}{\tau_f}\right)$ over time for case (h): $d(t) = t$ and $f(u, t) = 10^{-3} u^{-6} t^{-1}$. (a) Using solution result from coupled scheme. (b) Using solution result from split scheme.

then increased coupling over time is indicated since the solution decreases to 0 over time resulting in an increasing relative difference. Consequently for this case, the timescales metric indications do coincide with the relative difference results even though they do not match up with the log difference results.

With case (h) when $d(t) = t$ and $f(u, t) = 10^{-3} u^{-6} t^{-1}$, no stabilization is used to produce the results shown in Figure 34. In this case, the long-term behavior plateaus at zero for the coupled case and just below one for the split case. Both cases indicate strong coupling long-term. This behavior partially coincides with the corresponding log difference plot in Figure 8(b) since the values increase as time progresses; however, the coupling implied by the results in Figure 8(b) initially decreases and does not grow to be exceptionally strong in the time considered. If we incorporate the solution behavior shown in Figure 8(a), then the relative difference behaves similar to the log difference except the magnitudes over all are larger and the increase over time is faster since the solution decreases to zero. The coupling indications from

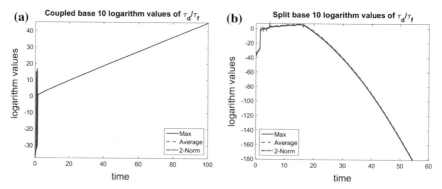

Fig. 35 Plots of $\log\left(\frac{\tau_d}{\tau_f}\right)$ over time for case (i): $d(t) = 2\exp(-t)$ and $f(u,t) = u^3 t^6$. (a) Using solution result from coupled scheme. (b) Using solution result from split scheme.

the relative difference correspond to the time scales metric results in this case even better.

When we consider case (i) with $d(t) = 2\exp(-t)$ and $f(u,t) = u^3 t^6$, we must utilize a different stabilization technique for each solution result. When using the coupled solution, the best results are obtained with the reset method and a tolerance of 1e-64 (see Figure 35(a)), while the best results when using the split solution are obtained with the reduction method and a tolerance of 1e-325 (see Figure 35(b)). For this case, the results using each solution are remarkably different from each other, but the same long-term coupling behavior is predicted by both. The results using the coupled solution start near zero (ignoring instabilities) and increase constantly away from zero, while the results using the split solution stay near zero (again ignoring instabilities) until almost $t = 20$ and then decrease almost linearly as time progresses. Both behaviors indicate strong coupling initially and a decrease in coupling long-term although the results using the split solution indicate strong coupling for a longer period of time. A comparison of the log difference results in Figure 9(b) shows similar end behavior for the coupling which should decrease as time increases. If we incorporate the solution behavior shown in Figure 9(a), the decrease in the solution is approximately the same rate as the decrease in the difference resulting in the relative difference being constant for large time. Consequently for this case, the metric results do not coincide with the relative difference indications although they do match up with the log difference indications.

With case (j) when $d(t) = 2\exp(-t)$ and $f(u,t) = 10^6 u^{-6} t^3$, no stabilization is required for either solution. Both plots in Figure 36 begin near zero and increase linearly away from zero as time increases indicating strong initial coupling which decreases with time. However, the opposite behavior is implied by the log difference results in Figure 10(b) since those values increase with time. Considering the solution behavior shown in Figure 10(a), the relative difference behavior is similar to the log difference behavior although with smaller magnitudes because of the increase in the

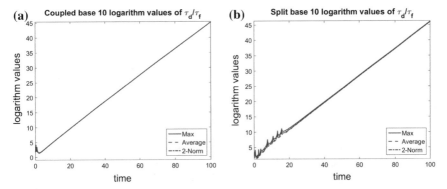

Fig. 36 Plots of $\log\left(\frac{\tau_d}{\tau_f}\right)$ over time for case (j): $d(t) = 2\exp(-t)$ and $f(u,t) = 10^6 u^{-6} t^3$. (a) Using solution result from coupled scheme. (b) Using solution result from split scheme.

solution. Consequently, the timescales metric does not show good predictions in this case.

The last nonlinear case (k), when $d(t) = 2\exp(-t)$ and $f(u,t) = u^3$, requires stabilization for both the coupled and split results shown in Figure 37. Just as with case (i), the reset method with a tolerance of 1e-64 produces the best results when using the coupled solution, and the reduction method with a tolerance of 1e-325 produces the best results when using the split solution. Again, the different solutions produce remarkably different results although the same coupling behavior is indicated by both: strong initial coupling that decreases with time. Just as with the previous case, the predicted coupling behavior does not match up with the corresponding log difference results (Figure 11(b)) which imply an increase in coupling as time progresses. If we consider the solution behavior shown in Figure 11(a), the relative difference is large initially since the solution starts near zero and then decreases slowly as time increases because the magnitude of the solution increases as time increases. Consequently, the time scales metric results match up better if we incorporate the solution behavior into the comparison.

4.3.3 Summary

In summary, it is interesting to note that stabilization is required with this metric for all cases when the power of u is greater than or equal to zero but not for any cases when the power of u is less than or equal to zero. This may contribute to the accuracy of the predictions with this metric since the predictions from those cases not requiring any form of stabilization generally match up better with the log difference implications, especially when considering the nonlinear test cases. The stabilization method does not necessarily decrease the efficacy of the metric although that is a possibility; it may simply be that this metric is not accurate for cases that produce

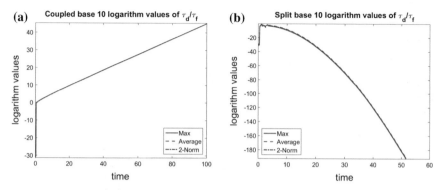

Fig. 37 Plots of $\log\left(\frac{\tau_d}{\tau_f}\right)$ over time for case (k): $d(t) = 2\exp(-t)$ and $f(u,t) = u^3$. (a) Using solution result from coupled scheme. (b) Using solution result from split scheme.

a solution which requires a stabilization method to be used in conjunction with this metric.

The timescales metric generally produces good predictions especially when considering the relative difference by incorporating the solution behavior although there are a few cases in which this metric fails to predict the correct coupling. Additionally, it is beneficial to see that the predictions made with coupled and split solutions match up even when the exact values do not.

5 Conclusions

We have investigated a number of potential metrics to use in determining whether the two physical components of a reaction–diffusion problem are coupled strongly. Three metrics were based on information found in the Jacobian matrix one would have through formulating the problem as a fully coupled system and solving with a Newton iteration. The first metric was based on the sizes of off-diagonal blocks of the Jacobian, and the second metric was based on the sizes of the diagonal blocks. The third metric was based on an error estimate of the solution formed from applying a Gauss–Seidel solution approach between the two system components. This error estimate was shown to be related to the first metric. The first two metrics could not be used for linear cases of the reaction function as they did not provide enough information to predict any changes in coupling strength. In many instances, these metrics did capture changes in coupling for more complex problems, although they were not consistently correct and even on occasion predicted the opposite trend of what was happening in the problem.

Despite these failings, the derivation of the first metric shows its relation to the nonlinear residuals of the problem subcomponents, and it is natural to consider this first metric as a part of a new metric that also uses the residual information. Future work will explore this idea in more detail. Overall, the metrics derived from the Jacobian matrix of the decoupled scheme seem less able to capture the coupling strengths in comparison with the timescales metric. The smoothness of these metrics may contribute to their failing to predict sharp coupling changes. For future study, we suggest investigating metrics derived not only from the Jacobian matrix but also from different sources.

The last metric considered was based on timescales of the component subproblems. Here, the timescales were each calculated, and the log of the ratio was investigated as a means to determine when the scales were close or far apart. This metric generally was better than the first in predicting coupling strength, but it also was not always correct. Again, there are other natural considerations for a metric like this to consider in future investigations, such as looking at differences rather than ratios or formulating a metric in terms of the difference between subproblem timescales and the dynamic timescale of the full problem.

The metric results shown in this paper are quite mixed in their coupling predictions. The metrics used here were many times approximated because of computational restrictions or for efficiency. These approximations may be one reason why they produce inaccurate results in some cases. In addition, the simplicity of the test cases may have resulted in some of the weak performance of the metrics. Generally as the complexity and nonlinearity of the test cases considered in this work increase, we see better performance of the metrics. In the linear cases, many of the metrics produced erroneous coupling predictions because of the simplicity of the diffusion function. Hence for future studies, the metrics should be tested with more complex cases, such as nonlinear diffusion–reaction and nonlinear advection–diffusion and greater spatial dimensions, to more completely evaluate their effectiveness.

Ultimately, we wish to construct a metric that is cheap to compute and can determine whether a decoupled scheme can be used in order to decrease computation time. Overall, the work in this paper reflects a step forward in developing algorithms that will allow efficient solution of multiphysics problems by adapting to the coupling strength of the components in the systems and solving individual problems when coupling is weak and monolithic problems when the coupling is strong. Such algorithms will become essential as computational scientists continue to move toward more complex, coupled simulations on ever larger computing platforms.

Acknowledgments The authors wish to thank Yekaterina Epshteyn for her guidance during the writing of this paper. This work was partially performed under the auspices of the US Department of Energy by Lawrence Livermore National Laboratory under contract DE-AC52-07NA27344. Lawrence Livermore National Security, LLC.

References

1. Estep, D., Ginting, V., Ropp, D., Shadid, J.N., Tavener, S.: An a posteriori-a priori analysis of multiscale operator splitting. SIAM J. Numer. Anal. **46**, 1116–1146 (2008). 10.1137/07068237X. http://portal.acm.org/citation.cfm?id=1362718.1362720

2. Hooper, R., Hopkins, M., Pawlowski, R., Carnes, B., Moffat, H.: Final report on LDRD project: Coupling strategies for multi-physics applications. Tech. Rep. SAND2007-7146, Sandia National Laboratory (2007)
3. Keyes, D.E., McInnes, L.C., Woodward, C., Gropp, W., Myra, E., Pernice, M., Bell, J., Brown, J., Clo, A., Connors, J., Constantinescu, E., Estep, D., Evans, K., Farhat, C., Hakim, A., Hammond, G., Hansen, G., Hill, J., Isaac, T., Jiao, X., Jordan, K., Kaushik, D., Kaxiras, E., Koniges, A., Lee, K., Lott, A., Lu, Q., Magerlein, J., Maxwell, R., McCourt, M., Mehl, M., Pawlowski, R., Randles, A.P., Reynolds, D., Rivire, B., Rde, U., Scheibe, T., Shadid, J., Sheehan, B., Shephard, M., Siegel, A., Smith, B., Tang, X., Wilson, C., Wohlmuth, B.: Multiphysics simulations: Challenges and opportunities. Int. J. High Perform. C. **27**(1), 4–83 (2013). 10.1177/1094342012468181
4. Knoll, D., Chacon, L., Margolin, L., Mousseau, V.: On balanced approximations for time integration of multiple time scale systems. J. Comput. Phys. **185**, 583–611 (2003)
5. Romero, L.: On the accuracy of operator-splitting methods for problems with multiple time scales. Tech. Rep. SAND2002-1448, Sandia National Laboratory (2002)
6. Ropp, D., Shadid, J., Ober, C.: Studies of the accuracy of time integration methods for reaction-diffusion equations. J. Comput. Phys. **194**, 544–574 (2004)
7. Ropp, D.L., Shadid, J.N.: Stability of operator splitting methods for systems with indefinite operators: Advection-diffusion-reaction systems. J. Comput. Phys. **228**, 3508–3516 (2009). 10.1016/j.jcp.2009.02.001. http://portal.acm.org/citation.cfm?id=1518324.1518454

Printed in the United States
By Bookmasters